The Scott Battle of 1817

First U.S. Defeat of the Seminole Wars

Dale Cox

2019

ISBN-13: 978-0-578-61757-2

(Previously published in large part as *The Scott Massacre of 1817*)

Visit the author online at:

twoeggflorida.com
scott1817.com

Old Kitchen Books
4523 Oak Grove Road
Bascom, Florida 32423

Cast forth lightning, and scatter them:
shoot out thine arrows, and destroy them.
Psalms 144: 6

This book is respectfully dedicated to
Linda Clark Smith

Table of Contents

Introduction

The Scott Battle of 1817, much like the earlier battle for Fort Mims and the later Battle of Little Bighorn, was a pivotal moment in American history. A crudely built wooden keelboat carrying a detachment of soldiers and eleven civilians was attacked by a force of Red Stick Creek, Seminole, Miccosukee, Yuchi, and Maroon or Black Seminole warriors on the Apalachicola River just below what was then the national boundary. Of the 50 or so people on board, only seven survived. Men and women were killed and scalped. Small children were picked up by their legs and killed by having their heads smashed against the sides of the boat. It was a gruesome retribution for an attack on a nearby Native American town and it forever changed the course of North American history.

The attack on Scott's command was direct and immediate retaliation for a raid by United States soldiers on the Lower Creek village of Fowltown. Following direct order from the administration of President James Monroe, troops from the Fourth and Seventh Regiments, U.S. Infantry, raided Fowltown on November 21, 1817. Their orders were to surround the village in the dark and take its charismatic chief, Neamathla (Eneah Emathla), hostage. Several warriors were killed and at least one woman died when the soldiers unleashed an indiscriminate volley on the men, women, and children of the town as they fled before their ranks. Furious over what they considered an unprovoked attack, the warriors on the Apalachicola fired on Scott's vessel without regard for the presence of women and children, repeating the example of the soldiers in blue at Fowltown. When the bloodshed was over, there was only one prisoner, a woman named Elizabeth Stewart. Every other woman, child, and all but six Scott's men were dead. Their bloody scalps were later found attached to a pole in the refugee camp established by the Fowltown survivors.

In justice to the warriors who took part in the attack, it must be pointed out that many of them were survivors of the Creek War of 1813-1814. They had seen their homes destroyed, towns burned, family members killed and everything they had ever known taken from them. An estimated 800 Creek warriors died at Horseshoe Bend alone and not even four years separated that devastating defeat from the attack on Scott's command. Other participants were Maroons or "Black Seminole" warriors, some of whom survived the destruction of the Fort at Prospect Bluff or "Negro Fort" just one year earlier. Their free settlement had been destroyed and 270 men, women and children killed in an explosion that U.S. officers called "horrible beyond description."

Scott's detachment, despite the presence of eleven women and children, was a legitimate military target. His vessel was a U.S. Army keelboat and a detachment of 40 American soldiers was onboard. The United States started the war at Fowltown, but it was the counterattack at what is now Chattahoochee, Florida, that brought the wrath of the Monroe Administration down on the Seminoles and their allies. In truth, the battle ignited a collision that had been building for a long time. The United States wanted Florida, believing it vital to the young country's defense. All that stood in the way was Spain and a handful of loosely organized Native American groups—and the Maroons. American authorities wanted the free blacks almost as bad as they wanted the Spanish lands themselves.

For the men, women and children of Scott's command, the collision brought only death and terror. For the United States, it was an atrocity that ended any reservations about launching an overwhelming campaign to destroy the Indian and Maroon towns across the line in Spanish Florida. President James Monroe and Secretary of War John C. Calhoun ordered Major General Andrew Jackson to the border with so many men that the warriors of the alliance that opposed him never really stood a chance. Jackson invaded Florida in compliance with direct, written instructions from the Monroe Administration and the history of the United States was forever changed.

Within four years of the Scott battle, Florida became a territory of the United States. Within seven years of the attack, most of the Seminoles and their allies were forced south into Central Florida. Within 20 years, the Trail of Tears was a reality. The battle near the head of the Apalachicola River had a lasting and dramatic impact. It was a Native American victory, but without it Florida might never have become part of the United States and the Seminole people might still reign supreme over the Land of Flowers.

The narrative that follows is the first book-length treatment of the Scott battle. It was the first U.S. defeat of the Seminole Wars and has long warranted such consideration. Since a slightly shorter version was first published in 2013, interest has surged in the story of the engagement. The historical significance of the battlefield on the banks of the Apalachicola is attracting greater attention. The local community raised a historical marker and the State of Florida in 2019 funded

Introduction

an eleven-stop interpretive trail at the site. Design of the interpretive panels is underway now and plans are underway for new archaeology on the battleground.

Many people assisted in the writing of this narrative and one is deserving of special recognition. Linda Clark Smith of the West Gadsden Historical Society offered much needed motivation and encouragement during the research and writing phases of this project. I am very grateful for her help and am pleased to dedicate the book in her honor.

The members of the West Gadsden Historical Society have long provided encouragement for my research and writing efforts and my appreciation is extended to them. The proceeds from the original version of the book benefited the WGHS in its ongoing effort to protect, preserve and interpret the history of Gadsden County, Florida. Thank you also to Savannah Brininstool, the editor of the original edition. She is a friend and served in the Middle East as a dedicated soldier of the District of Columbia National Guard.

Recognition is also due to the City of Chattahoochee, the Chattahoochee Rotary Club and Kathy S. Foster of the now defunct Twin-City News for their encouragement. I have much fondness for Chattahoochee and its citizens. Brian Rucker of Pensacola State College strongly encouraged me to delve deeper into the history of the First Seminole War. The late Judge E.W. Carswell of Chipley provided much documentation during his lifetime and I always think of him as I complete a new book. My uncles, Thomas Cox and Leroy Cox, were long-time residents of Chattahoochee and used to take me exploring the bluffs, ravines and riverbanks searching for long-forgotten historic sites. I will never forget them. Thank you to my sons, William Cox and Alan Cox, for their assistance and encouragement. My special appreciation also is extended to my mother, Pearl Cox, for reading through the first draft of the manuscript.

Appreciation must be extended to the staff members of the following institutions: Florida State Archives, Georgia State Archives, Alabama State Archives, John C. Pace Library of the University of West Florida, Florida History Library at the University of Florida, Georgia Historical Society in Savannah, Bradley Library in Columbus (GA), Willard Library in Evansville (IN), Fort Smith Public Library in Fort Smith (AR), Richland Library in Columbia (SC), Fort Smith National Historic Site (AR), Gulf Islands National Seashore (FL), San Marcos de Apalache Historic State Park (FL), Fort Gadsden Historic Site (FL), Fort Gibson State Historic Site (OK), the Library of Congress and the National Archives

Many others too numerous to name have assisted me in one way or another through the years in the writing of this book. Thank you to all.

Special thanks also to my friend, business partner, and fellow dog parent Rachael Conrad, who listens to be drone on for hours about the Trail of Tears, Creek and Seminole Wars, and other topics. Without her help I would be eternally lost in the weeds of both my work and my mind.

May God bless and keep you.

Dale Cox
November 25, 2019

The Scott Battle of 1817

First U.S. Defeat of the Seminole Wars

Chapter One

The Year without a Summer

On November 30, 1817, a boat carrying soldiers and supplies, as well as seven women and four children, rounded the sharp bend of the Apalachicola River where the city of Chattahoochee stands today. The current was strong and despite the efforts of the men pulling at the oars, the boat was pushed close to the east bank of the river. Lieutenant Richard W. Scott of the 7th U.S. Infantry regiment commanded the vessel and had been warned that he might face an attack from Red Stick Creeks and their Seminole Indian allies before he reached the confluence of the Flint and Chattahoochee Rivers.

Scott and his soldiers, half of whom were sick and unarmed, had kept careful watch but so far the attack they expected had not materialized. By the morning of the 30th, the boat was less than two miles from the confluence and only twelve miles from the safety of Fort Scott. Fighting to keep the boat moving forward as it was pushed near the east bank by the current, the lieutenant and his men never saw the warriors who were waiting there for them.

The shoreline erupted with a sheet of flame as lead balls exploded from hundreds of rifles and muskets, all aimed at Lieutenant Scott and his command. The bloodbath that followed forever changed American history. The U.S. Army sustained its first defeat of the Seminole Wars, a series of conflicts that would plague the nation for four decades to come. In their victory, the Seminoles and their Red Stick allies assured their defeat in the greater war. Andrew Jackson would be sent to invade foreign soil, clearly demonstrating to Spain that it could not hope to hold Florida. The old colony became a U.S. territory just four years later.

The autumn of 1817 came during a time of tumult and chaos in the world. The spell started in 1811 when a brilliant comet appeared in the skies. Massive earthquakes shook the New Madrid Fault that same year, so rattling the United States that politicians took to the streets in Washington, D.C., believing that the capital was somehow under attack. The natural phenomena at least partially helped spark a time of great disturbance in the course of world history. The once mighty Creek Nation was shattered at Horseshoe Bend in March 1814 and that same year Andrew Jackson smashed the British at the Battle of New Orleans. Napoleon fell, rose, and fell again. And just as it seemed as if peace might return to the world, the volcano Mount Tambora erupted in the Pacific.

Ash from the explosion drifted high into the atmosphere and slowly spread around the world. So much of it infiltrated the sky that the year of 1816 was remembered for decades to come as the "Year Without a Summer." Strange weather destroyed the corn crops in New England and Europe. Hundreds of thousands died, and food riots shattered the peace of Switzerland. In the United States, the light of day glowed in a strange golden or orange color and unseasonable cold persisted far longer than anyone had ever seen. The unusual and deadly weather continued for two years to come.

In Florida during the summer of 1816, U.S. forces destroyed the so-called "Negro Fort" on the Apalachicola River. A heated cannonball passed through the open door of a gunpowder magazine, instantly ignited an explosion that shook the ground as far away as Pensacola. In a blinding flash, 270 of the estimated 320 men, women and children in the fort perished.

With them was destroyed or captured a vast armament of muskets, carbines, swords, gun flints, cannon, powder and other military supplies that had been left behind by the British at the end of the War of 1812. The totality of the disaster so stunned the Lower Creek and Seminole Indians of Florida, Southwest Georgia and South Alabama that they did not resist the movement of the American troops as they marched back up the river to Georgia. Returning to a temporary outpost called Camp Crawford on the lower Flint River, they began building a more permanent fort with squared log buildings and comfortable quarters. The new post was called Fort Scott, after General Winfield T. Scott, and was described as "elegant" by one inspector who reviewed the work in progress.

The fort was still incomplete in December 1816 when orders came for Lieutenant Colonel Duncan Lamont Clinch and his men from the 4th U.S. Infantry to evacuate it. Leaving the buildings and supplies stored there in the care of George Perryman, the mestizo brother of the Lower Creek chiefs William and Ben Perryman, the troops evacuated Fort Scott and headed north to Fort Gaines and then Fort Mitchell.

The move gave new confidence to the refugee Red Stick fighters concentrated near the confluence of the Flint and Chattahoochee Rivers. They had fled to the region after their bloody defeat at the Battle of Horseshoe Bend seeking and receiving resupply from the British who had built two forts on the Apalachicola. Irate over the destruction of their depot at the "Negro Fort," they now exacted their revenge on the unfinished and unguarded buildings of Fort Scott. Appearing almost as soon as the soldiers departed, they threatened Perryman and drove him off from the place. He was only able to secure his family and a few personal possessions before fleeing the scene in a canoe. The supplies left at the fort were ransacked and stolen and fire was set to the log buildings. George Perryman saw at least three buildings burning as he fled. The Indians, he said, were "in numbers."[1]

It is generally believed, and probably accurately so, that the Fowltown warriors were among those who raided Fort Scott. The town had joined with the Yuchis in "taking the talk" of the Prophet Josiah Francis and his followers during the Creek War of 1813-1814.

Francis, it was said by his followers, could communicate with the water spirits. He was often seen walking down into a flowing stream or river from which he would not emerge for many hours. A practitioner of the Nativistic religion of the Shawnee Prophet Tenskwatawa, Francis taught the converts who gathered around him that the Indians should separate themselves from the whites, that accommodationist chiefs should be overthrown and that the Creeks should live in peace with all people, but subject to none. No more Native American land was to be given up.

Such talk resonated with the head chief of Fowltown (called Tutalosi Talofa or "Chicken Town" in the Hitchiti tongue of the Lower Creeks). His name or title was Eneah Emathla and he was charismatic leader and courageous man. The whites, who were never very good with Indian names, called him everything from E-nee-hee-maut-by to Eneamathla before finally settling on the name Neamathla.

One contemporary said that Neamathla was able to command his warriors with a mere look. He agreed with Josiah Francis that the Creeks should no longer submit to the expansionist desires of the whites and led his warriors to join the Prophet's force at Holy Ground on the Alabama River. The Tutalosi warriors, however, were cornered and defeated at the Battle of Uchee in what is now Russell County, Alabama, by an overwhelming party of Cowetas led by the U.S.-allied chief William McIntosh.

Blocked in his objective of joining the Red Sticks in Alabama and fearing that the women and children of his town would be subjected to follow-up attack by McIntosh's warriors, Neamathla abandoned his town on Kinchafoonee Creek near present day Albany, Georgia, and withdrew down the Flint River into the deep wilderness near the Florida border. There he resettled his people and established a new town, also called Tutalosi Talofa or Fowltown.

Neamathla and his warriors quickly allied themselves with the British, who appeared on the Apalachicola at about the same time. Joining the auxiliary force being raised by Lieutenant Colonel Edward Nicolls, they were given arms, ammunition and a drum for their town. The chief himself was presented with a military uniform coat of scarlet color and a letter signed by Captain Robert Spencer testifying that Neamathla was a loyal and good friend of Great Britain.

Neamathla was present at the British outpost near the confluence of the Chattahoochee and Flint Rivers on March 10, 1815, when a large gathering of Creek and Seminole chiefs signed a written appeal to the Prince Regent in London for help in enforcing the provisions of the Treaty of Ghent. The treaty had finally ended the War of 1812 and required that all combatants returned to their prewar land holdings. Colonel Nicolls and other British officers on the Apalachicola believed that the Creeks were covered under the terms of the treaty as their war with the United States had been a subsidiary part of the larger conflict. The United States disagreed and maintained that the Creeks had already entered into a separate treaty and therefore were not covered under the Treaty of Ghent.

That agreement, called the Treaty of Fort Jackson, was negotiated by Andrew Jackson in August 1814. Signed by many Creek leaders, it was in effect a Native American surrender. In exchange for peace, the chiefs agreed to Jackson's terms and ceded to the United States a vast stretch of territory including 23,000,000 acres. The largest swath of this cession was along the Florida border and, unfortunately for Neamathla and his followers, it included the land on which they had established their new village.

The presence of Fowltown and other Lower Creek villages on the Treaty lands was not an immediate concern to anyone. Few whites were brave enough to settle in the new territory and the Indians for the most part just wanted to be left alone. The burning of Fort Scott by Red Stick warriors in January 1817 changed this equation, however, and events on the frontier soon spiraled out of control. A flurry of letters and reports traveled back and forth between the frontier and both the War Department in Washington, D.C., and the headquarters of Major General Andrew Jackson in Nashville, Tennessee.

The blame for the burning of Fort Scott quickly focused on Neamathla and his warriors at Fowltown. Although they often debated the proper policy to pursue with regard to the Indians of the Southeast, in this case U.S. officials reached an almost unanimous conclusion: Neamathla would have to remove himself and his people from the ceded lands. The chief, however, believed the land was his and that he was "directed by the powers above to defend it." The stage was set for war and the conflict was not long in coming.

---

[1] Lt. Richard Sands to Maj. Gen. Edmund P. Gaines, February 2, 1817.

Chapter Two

The Rebuilding of Fort Scott

Along the Florida border, the tension escalated. On February 24, 1817, George Perryman wrote to Lieutenant Richard Sands at Fort Gaines warning him that the Fowltown warriors were engaged in predatory raids against Georgia's frontier settlements:

There was a friend of mine not long since in the Fowltown, on Flint, and he saw many horses, cattle, and hogs, that had come immediately from the State of Georgia; and they are bringing them away continually. They speak in the most contemptuous manner of the Americans, and threaten to have satisfaction for what has been done – meaning the destruction of the negro fort.[1]

The letter went on to warn that an army of Seminole, Red Stick Creek and African (or Black Seminole) warriors was being formed. Their leader was the Alachua Seminole chief Boleck (Bowleg), who saw his town burned and his brother killed by U.S. troops that invaded East Florida in 1812:

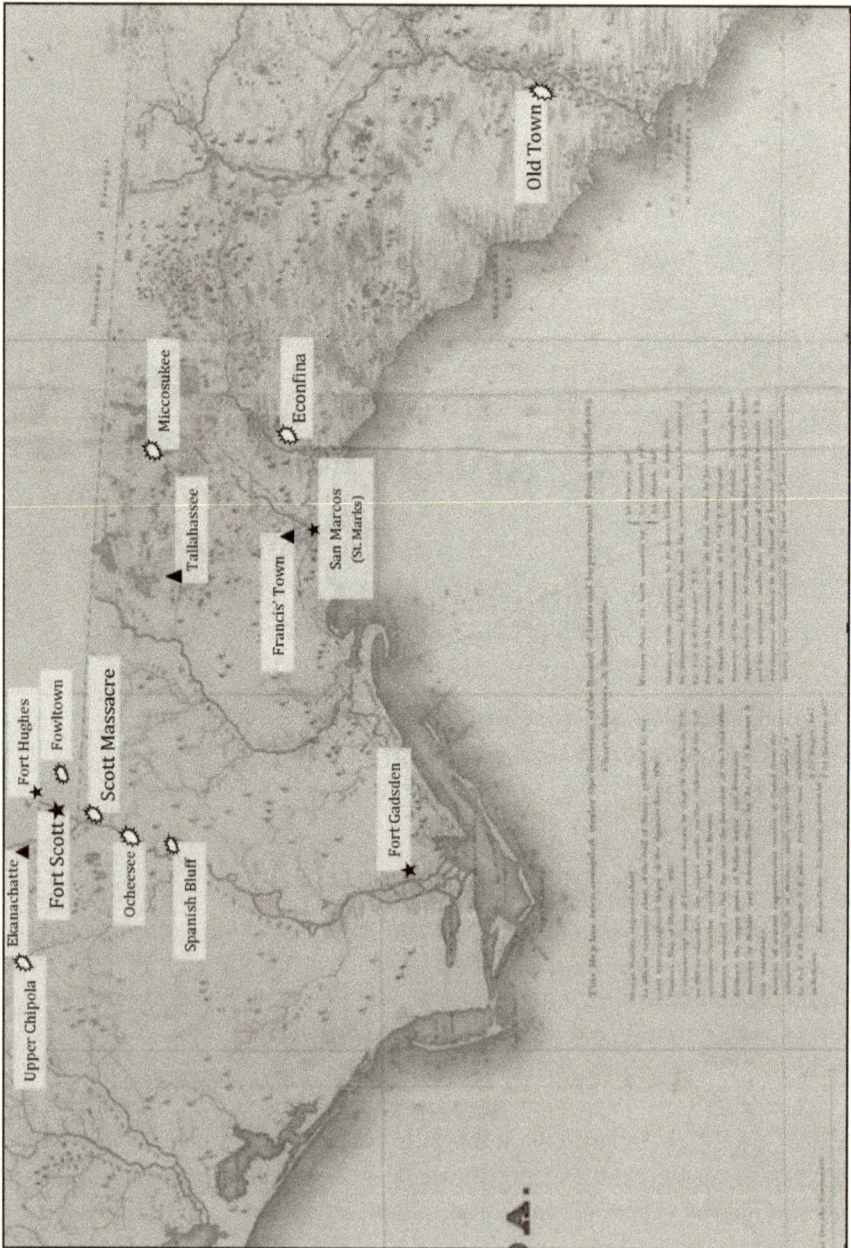

...They say they are in a complete fix for fighting, and wish an engagement with the Americans, or McIntosh's troops; they would let them know they had something more to do than they had at Appalachicola. They have chosen [Boleck] for their head, and nominated him king, and pay him all kind of monarchical respect, almost to idolatry, keeping a picket guard at a distance of five miles. They have a number of the likeliest American horses; but there are one or two chiefs who are not of the choir. Kenhijah, the Mickasuky chief, is one that is an exception.[2]

Perryman's warning was punctuated on the same day when Seminole warriors attacked the Garrett home in western Camden County not far from the Okefenokee Swamp. The bloody incident was reported to General Gaines by Archibald Clarke, the St. Mary's Intendant, on February 26th:

On the 24th instant, the house of a Mr. Garrett, residing in the upper part of this county, near the boundary of Wayne county, was attacked during his absence, near the middle of the day, by this party, consisting of about fifteen, who shot Mrs. Garrett in two places, and then despatched her by stabbing and scalping. Her two children, one about three years, the other two months old, were also murdered, and the eldest scalped; the house was then plundered of every article of value, and set on fire. A young man in this neighborhood, hearing the report of guns, went immediately towards the house, where he discovered the murdered family.[3]

The youth who discovered the murders extinguished the flames before they could consume the house and then alerted the neighborhood. Archibald Clarke's mills were nearby and his workmen, along with several other neighbors, set out in pursuit of the war party. They failed to catch the warriors and ended the chase rather than follow the Seminoles deep into the wilderness without enough arms and equipment.[4]

Clarke's report raged with frustration as he pleaded with Gaines for help:

On this open, extensive, and entirely unprotected frontier, the poor and innocent inhabitants have ever been exposed to these calamities. Representation after Representation to the several Governors of this State, of cruel and unprovoked murders in this quarter by the Indians, have been made. A momentary disposition was manifested to afford relief; but little time, however, would elapse before the alarm would subside, and the subject never more thought of again until revived by an occurrence such as I have just related.[5]

Clarke asked that the general order a detachment of troops to the head of the St. Mary's for the protection of the settlers and their cattle herds. Before Gaines could even receive the intendant's letter, another communication came up the Apalachicola from the other side of the border.

A Scottish trader named Alexander Arbuthnot had arrived on the coast of the Big Bend of Florida. A shrewd businessman who hoped to open a profitable trade with the Seminoles and refugee Red Sticks, he seems to have taken a real interest in the affairs of his customers. On March 3, 1817, he attempted to open a correspondence with the U.S. Army officer commanding at Fort Gaines:

The Lower Creeks seem to wish to live peaceably and quietly, and in good friendship with the others; but there are some designing and evil-minded persons, self-interested, who are endeavoring to create quarrels between the Upper and Lower Creek Indians, contrary to their interest, their happiness, and welfare. Such people belong to no nation, and ought not to be countenanced by any Government. The head chiefs request I will inquire of you why American settlers are descending the Chattahoochee, driving the poor Indian from his habitation, and taking possession of his home and cultivated fields.[6]

The American settlers mentioned by Arbuthnot, who dated his letter from "Ochlochnee Sound," were coming down to settle on lands near Fort Gaines that had been surrendered by the Creek Nation under the terms of the Treaty of Fort Jackson. The United States had no interest in discussing the ceded territory, even though many of the chiefs from along the border had not been party to the treaty.

Any sympathy or consideration the trader might have received from U.S. officers was eliminated when he appealed the case of Peter McQueen, one of the principal Red Stick leaders in the Creek War of 1813-1814. McQueen had lost a number of slaves when he was forced to flee to Florida to escape the American armies. He now sought Arbuthnot's help in having them returned to him:

...The American headmen and officers that were accustomed to live near him can testify to his civility and good fellowship with them; and there are none of them, he is convinced, that would not serve him, if in their power. As he owes nothing, nor ever took any person's property, none have a right to retain his; and he hopes that, through your influence, those persons now holding his negroes will be induced to give them up.[7]

The Americans were not impressed with McQueen's protestations of friendship and Arbuthnot's association with him permanently damaged the Scotsman's reputation in their eyes. He emphasized in his letter that he held no official position and hoped that his attempt to intervene on behalf of the Indians would not be regarded as undue interference on his part. In fact, that is exactly the conclusion they reached about it and him.

Their opinion of the new arrival below the border darkened even more when William Perryman, brother of George and chief of the town of Tellmochesses on the lower Chattahoochee River, arrived at Fort Gaines with intelligence about the activities of Arbuthnot and McQueen:

Yesterday, William Perryman, accompanied by two of the lower chiefs, arrived here. He informs me that McQueen, the chief mentioned in one of the enclosed letters, is at present one of the heads of the hostiles; that they are anxious for war, and have lately murdered a woman and two children.

He likewise says he expects the news in George Perryman's letter is true; for there are talks going through the towns that the English are to be at Ochlochnee river in three months.

I have sent an Indian runner to Ochlochnee to ascertain what preparations the hostiles are making.[8]

The murders referred to by Perryman were those of Mrs. Garrett and her children near the St. Mary's. The chief's view of the situation was worthy of deep consideration by American officials as he had lived just below the Florida border since the time of the American Revolution. He had saved the life of U.S. Commissioner of Limits Andrew Ellicott during the survey of the boundary between the United States and Spanish Florida in 1799 and had opposed the designs of William Augustus Bowles during the same era. His father, Thomas Perryman, had been the head chief of the Seminole Nation until the time of his death about two years earlier.

The new U.S. Indian Agent, former Georgia governor David B. Mitchell, took a somewhat more moderate view of the situation below the border. In a letter to the Secretary of War two weeks after Perryman's visit to Fort Gaines, he reported the receipt of intelligence that the Red Sticks had resumed their war dances. The Garrett murders, he reported, were in retaliation for the murder of Indians below the border by whites from Georgia and he expected "no further bad consequences from it." The real problem, he felt, was the decision to evacuate Fort Scott, which he still called Camp Crawford. The departure of the troops, he noted, was blamed on fear by the Indians and had reinvigorated them. The presence of Arbuthnot on the coast was not helping:

...I have received information from other persons at or near Fort Gaines that a British agent is now among these hostile Indians, and that he has been sending insolent messages to the friendly Indians and white men settled above the Spanish line: he is also charged with stimulating the Indians to their present hostile aspect; but whether he is an acknowledged agent of any foreign Power, or a mere adventurer, I do not pretend to determine, but am disposed to believe him the latter; but, be that as it may, and let the hostile disposition of the Indians proceed from what it may, a moderate regular force stationed at Camp Crawford, or any other suitable position in that quarter, will, I am confident, keep all quiet; and, without it, some serious mischief will result.[9]

General Gaines agreed with Mitchell's assessment and notified officials in Georgia that while he lacked the authority to return the battalion from the 4th U.S. Infantry regiment to Fort Scott, he would seek permission from the War Department to order the men back. Should he be refused such authority or should the soldiers be too far advanced on their march to the Alabama River, he assured Governor William Rabun that he would order a company or two of artillery to the Flint River to occupy the fort and put the works in order.

The latter option was resorted to and on April 27, 1817, Captain Samuel Donoho left Charleston with his company from the 4th U.S. Artillery:

Charleston, April 28. – A company of U.S. Artillery, under command of Capt. Donoho, marched from this city yesterday for Fort Scott, near the boundary line between East Florida and Georgia. They are to act against the hostile Indians who have recently been making depredations.[10]

Donoho's unit, which appears to have been ordered to the frontier as "red-legged infantry" (artillerymen temporarily assigned to infantry duty), was relatively small. Company returns for 1817-1818 show that the company normally averaged 28-32 men counting Donoho and his first lieutenant, Milo Johnson. Sending so few men to such a remote post in a time of increasing hostility and rumor of war was extremely dangerous, but they undertook the mission without recorded complaint.[11]

The artillerymen reached Augusta, Georgia, in early May, pausing there briefly before continuing their march on to Milledgeville and eventually the frontier. Their movement was long and slow, and it was not until the first part of June that the solders finally reached Fort Scott.[12]

The fort proved to be much more severely damaged than anyone had expected. The fires set by the Red Sticks had destroyed not only the barracks, but the officers' quarters and most of the other structures of the post. The artillerymen started work cutting logs and dragging them to the ruined compound, but the challenge before them was daunting. It would take until December for the fort to be declared complete. Donoho's small company remained alone and exposed in the wilderness for roughly one month until reinforcements from the 7th U.S. Infantry arrived in July under Brevet Major David E. Twiggs. Even with the arrival of the major's company, however, the total strength of the command at the fort numbered only 116 men.[13]

---

[1] George Perryman to Lt. R. Sands, February 24, 1817, *American State Papers*, Indian Affairs, Volume II, p. 155.

[2] *Ibid.*

[3] Archibald Clarke to Maj. Gen. E.P. Gaines, February 26, 1817, *American State Papers*, Indian Affairs, Volume II, p. 155.

[4] *Ibid.*

[5] *Ibid.*

[6] Alexander Arbuthnot to the Commanding Officer at Fort Gaines, March 3, 1817, *American State Papers*, Indian Affairs, Volume II, p. 155.

[7] *Ibid.*

[8] Lt. Richard M. Sands to Col. William King, March 15, 1817, *American State Papers*, Indian Affairs, Volume II, p. 156.

[9] David B. Mitchell to the Secretary of War, March 30, 1817, *American State Papers*, Indian Affairs, Volume II, pp. 156-157.

[10] *Vermont Reporter*, May 20, 1817, p. 3.

[11] Post Returns, Fort Scott, Georgia, 1817-1821, National Archives.

[12] *Poulson's American Daily Advertiser,* July 11, 1817, p. 3.

[13] Post Returns, Fort Scott, Georgia, 1817-1821, National Archives.

Chapter Three

Escalation on the Border

Neamathla was not pleased with the renewed presence of U.S. soldiers less than twelve miles from his village. He and his people had worked hard to reestablish themselves along Four Mile Creek south of what is now Bainbridge after they had evacuated their old town site near Albany. Once again settled into comfortable homes with their livestock herds replenished, he and the people of Fowltown had come down the Flint River to escape the whites and their allies, Big Warrior and William McIntosh. They also appear to have had a hereditary claim to the land on which they settled.

Thirty years previous, during the American Revolution, a British force had passed through the area on its way to reinforce St. Augustine against expected attack by Patriot forces from Georgia. Great Britain had claimed Florida as a prize at the end of the French and Indian War in 1763 and the old Spanish city served as capital of the Royal colony of East Florida. There were constant attacks back and forth across the border as British and Loyalist forces battled Patriot troops for control of the Georgia coast. In response to a threatened invasion, the governor in St. Augustine sent out a call for help from the West Florida capital of Pensacola. British officers there responded by sending an official party east on the old

Pensacola to St. Augustine trail with orders to enlist Indian forces in the region to fight for the British.

Thomas and William Perryman, who lived on the lower Chattahoochee, were already actively engaged as partisan fighters for the British. The 1778 expedition resulted in the warriors of Ekanachatte, Tomatley, Pucknawhitla (Burges's Town) and Oklafunee also joining the cause. The latter two towns were on the Flint at Bainbridge and where Four Mile Creek flowed into the river respectively.

Burges's Town took its name from James Burges (or Burgess), an 18th century trader who married the daughter of the chief of Pucknawhitla and operated his primary store there. He had a secondary store (and another wife) at Tomatley, a town just below the forks of the Chattahoochee and Flint Rivers in what is now Jackson County, Florida. Another trader from down the Apalachicola named John Mealy met the British at Burges's Town with horses to assist in their operations and the growing command then continued its journey along the Pensacola to St. Augustine road.

As the expedition advanced, the cartographer Joseph Purcell prepared a detailed map of the route, with itineraries. This map, the original of which is preserved today at the Library of Congress, showed that at about 7 ½ miles below Burges's or Bainbridge, the old road split into two branches. The main trail continued on across the Ochlockonee River to Tallahassee Talofa and Miccosukee, early Seminole towns in Leon County, Florida. A secondary trail, called the Harmonia Path, branched more to the north and in three miles crossed a stream labeled "Tootoloosa-Hopunga Creek" before leading on to the villages of Ochlockonee and Miccosukee. It rejoined the main trail at the latter town.[1]

"Tootoloosa" as given on the map, of course, is identical to the Tutalosi of later times that referred to the inhabitants of Tutalosi Talofa or Fowltown. The term "Hopunga," in the Hitchiti dialect, means something akin to "broken up" or "destroyed." The presence of a Tootoloosa-Hopunga Creek 10 ½ miles east of Bainbridge, then, indicated that by 1778 a Tutalosi town had already existed in that vicinity and been either abandoned or destroyed.

This fact clearly explains why Neamathla and his followers believed they had a long-standing claim to the land west of the Flint along the Florida border, even though they had established their new Fowltown there only two or three years earlier. It is reasonable to speculate that the ancestors of the Fowltown people – and probably some of their living members – had lived near Tootoloosa-Hopunga Creek before moving up the Flint River to establish the town on the Kinchafoonee where they resided until the Creek War of 1813-1814.

In addition, the chief clearly believed that because he had not been a party to the negotiations of the Treaty of Fort Jackson, he was not bound by that agreement. In short, Neamathla believed that from both a hereditary and a legal standpoint, the land below the Flint River was his and he meant to keep it.

As the backbreaking work of rebuilding Fort Scott continued on the lower Flint, the U.S. government took a step that greatly intensified the anger of the Red Sticks and Seminoles along the Florida border. Major Twiggs was ordered to convene a council at Miccosukee, the principal Seminole town west of the Suwannee, and demand that the murderers of the Garrett family be surrendered to the whites. Twiggs sent the interpreter Gregory to read this demand to an assembly of chiefs and warriors at Miccosukee on September 6, 1817:

...[T]hose who were present said they had never heard of Indians being given up to be punished by the whites; that they had heard of their being sometimes killed by themselves for offences committed, but seemed to think that giving them up was out of the question, but said they would have a meeting, and would answer the letter in a few days. As they have not done so, I think but one construction can be put on their conduct.[2]

Cappachimico, the chief of Miccosukee, actually did not delay in sending his response. It was taken by a runner to Fort Hawkins instead of Fort Scott, probably because the Seminoles were unclear as to where they should send their reply. Since Fort Hawkins was a long-established post that was familiar to them, they sent it to the officers there.

The council at Miccosukee appears to have been tense. Although the chiefs and older warriors heard Gregory with politeness, as was their custom, the younger warriors were outraged by the demand that Indians be turned over to the whites for punishment:

...The young men seemed to dislike the communication very much, and when Gregory was about leaving the town he offered his hand to an Indian, who held out his with a knife in it, and refused to shake hands with him; he staid so short a time among them that it was impossible for him to give much information respecting them.[3]

Cappachimico's response to the demand came in writing, indicating that he was probably assisted in writing it by Alexander Arbuthnot, although it is also possible that the letter was penned by either Edmund Doyle or William Hambly. Long-time employees of Forbes and Company who lived and worked on the Apalachicola River, both men were well-known to the chief.

In his response, the Seminole leader indicated that it would not be possible to turn over the men accused of murdering Mrs. Garrett and her two children to the whites for punishment. He also attempted to explain that the Indians actually had much more to complain about than the whites, pointing out that a number of his people had been killed by Georgians since the end of the War of 1812:

...The whites first began, and there is nothing said about that, but great complaint made about what the Indians do. This is now three years since the white people killed three Indians; since that they have killed three other Indians, and taken their horses and what they had; and this summer they killed three more, and very lately they killed one more. We sent word to the white people that these murders were done, and the answer was that they were people that were outlaws, and we ought to go and kill them. The white people killed our people first, and the Indians then took satisfaction. There are yet three men that the red people have never taken satisfaction for.[4]

With regard to the Garrett murders, the chief attempted to explain that the woman's husband was believed by the Indians to have been involved in crimes against them:

There were some of our young men out hunting, and they were killed. Others went to take satisfaction, and the kettle of the ones that were killed was found in the house where the woman and two children were killed; and they supposed it had been her husband who had killed the Indians, and took their satisfaction there. We are accused of killing up Americans, and so on; but since the word was sent to us that peace was made, we stay steady at home and meddle with no person.[5]

Cappachimico's response is one of America's most intriguing historical documents. It presents the view of the principal Seminole leaders on the eve of the outbreak of a series of wars that would continue for the next forty years. In fact, until recent years, the Seminole conflict, in particular the Second Seminole War (1835-1842), was regarded as the longest war in U.S. history. By the time the wars ended, thousands of Indians had been forcibly removed from their homes and sent west to present-day Oklahoma, hundreds of U.S. soldiers had been killed and the last remaining Seminoles had been pushed far down into the Florida peninsula. In the end, though, they never surrendered.

As tensions grew between the U.S. government and the powerful Miccosukees, Neamathla issued a firm warning to Major Twiggs at Fort Scott. The soldiers there were cutting timber for use in rebuilding the fort and the chief, in no uncertain terms, warned the major not to let his men cross to the opposite side of the Flint River:

By a letter from Major Twiggs, the commandant at Fort Scott, I learn that he had been warned some weeks past by the principal chief of Fowltown (fifteen miles above the fort, and twenty above the national boundary) not to cut another stick on the east side of Flint river; adding that the land was his, and he was directed by the Powers above to protect and defend it, and should do so; and it would be seen that talking could not frighten him. Major Twiggs adds, he had not seen the chief or any of his people since he made this threat.[6]

The warning from Neamathla was, in effect, a drawing of a line in the sand. The Indians meant to defend the lands south and east of the Flint River and any attempt by the whites to cross or cut timber there would be met by armed resistance. The line had been drawn and it was now up to General Gaines and his superiors to decide whether or not to cross it.

The general's response was predictable. He ordered the movement of the 4th and 7th U.S. Infantry regiments to Fort Scott. Gaines appears to have hoped that the appearance of the large force alone would sufficiently intimidate the Red Sticks and Seminoles into backing down and meeting the U.S. demands for the murderers of the Garrett family. President James Monroe agreed:

These papers have been submitted to the President, and I am instructed by him to inform you that he approves of the movement of the troops from Fort Montgomery to Fort Scott. The appearance of this additional force, he flatters himself, will at least have the effect of restraining the Seminoles from committing further depredations, and, perhaps, of inducing them to make reparation for the murders which they have committed. Should they, however, persevere in their refusal to make such reparation, it is the wish of the President that you should not on that account pass the line and make an attack upon them within the limits of Florida, until you shall have received instructions from this Department.[7]

The letter from acting Secretary of War George Graham went on, however, to authorize General Gaines to move against the Fowltown Indians and any others who were still living on lands ceded to the United States by the Treaty of Fort Jackson. In fact, Monroe and Graham even approved the taking of hostages by U.S. forces if Gaines thought it advisable:

You are authorized to remove the Indians still remaining on the lands ceded by the treaty made by General Jackson with the Creeks; and, in doing so, it may be proper to retain some of them as hostages until reparation may have been made for the depredations which have been committed. McIntosh and the other chiefs of the Creek nation, who were here some time since, expressed then, decidedly, their unwillingness to permit any of the hostile Indians to return to their nation.[8]

The idea of a United States President authorizing the taking of hostages is almost unfathomable today, yet Monroe was clearly willing to stoop to this level if it would force the Seminoles to return stolen cattle and surrender the murderers of the Garrett family. Gaines, to his credit, never attempted to act on the authorization.

---

[1] Stuart-Purcell Map of 1778, Cartographic Division, Library of Congress.

[2] Maj. David E. Twiggs to Maj. Gen. Edmund P. Gaines, September 17, 1817, *American State Papers*, Indian Affairs, Volume II, p. 158.

[3] *Ibid.*

[4] Cappachimico to Commanding Officer at Fort Hawkins, September 18, 1817, *American State Papers*, Indian Affairs, Volume II, p. 159.

[5] *Ibid.*

[6] Maj. Gen. Edmund P. Gaines to the Secretary of War, October 1, 1817, *American State Papers*, Indian Affairs, Volume II, pp. 158-159.

[7] George Graham, Acting Secretary of War, to Maj. Gen. Edmund P. Gaines, October 30, 1817, *American State Papers*, Indian Affairs, Volume II, p. 159.

[8] *Ibid.*

Chapter Four

The Battle of Fowltown

THE MOVEMENT OF THE 4TH AND 7TH INFANTRY REGIMENTS, which comprised the First Brigade of the U.S. Army, from Camps Montgomery and Montpelier in Alabama to Fort Scott was a daunting task. There were no direct roads for the soldiers to follow and the necessary transport of supplies and equipment for so large a force to a spot so deep in the wilderness was almost inconceivable in that day and age.

As General Gaines ordered his men to begin their more than 250 mile march, he did everything possible to prepare for their arrival at their new post and to supply their movement as they advanced. In hopes of shortening the distance, the general ordered that a new road be cut overland through the woods to Fort Gaines on the Chattahoochee. This road, which was more a rough path cleared by ax-swinging soldiers, eliminated the need for the force to march all the way up the Federal Road to Fort Mitchell and then travel by water down the Chattahoochee either to Fort Gaines, and from there by land to Fort Scott, or all the way down to the confluence and then up the Flint to the fort.

As the main force prepared to begin is overland march, the general ordered Brevet Major Peter Muhlenberg of the 4th Infantry to proceed to Fort Scott by water from Camp Montgomery near Tensaw, Alabama, with vital supplies for the soldiers:

You will embark with the detachment assigned to you, on board the transports now at the landing, as soon as they shall be ready for your reception, and repair to Fort Scott upon the Flint River. The vessels are to receive in addition to the ordnance stores and baggage of the troops, such contractors stores as Mr. R. Tankersley, the principal agent at Mobile, may put on board to complete the cargo of each vessel, provided however, the said stores shall consist principally of salted pork, together with vinegar, soap and candles to be delivered to the contractors agent at Fort Scott, who will receipt for the same as a part of the supply ordered to be forwarded in charge of Lt. Scott, and which it appears he was compelled for want of room on board his transports to leave at Mobile.[1]

The orders to Major Muhlenberg indicate that 1st Lieutenant Richard W. Scott of the 7th had already been ordered forward by water with a supply of provisions. A native of Virginia and veteran of the War of 1812, Scott was a lieutenant in the company of Major David E. Twiggs, which had arrived at Fort Scott in July. His experience in navigating the Apalachicola River would prove fatal to both him and the men of his command before the end of November.

Muhlenberg was cautioned by Gaines to be extremely alert while moving from Mobile Bay to Fort Scott, not just because of the risk of Indian attack but because of the possibility of an encounter with pirates as well:

The unfriendly character of the Seminola Indians and other persons inhabiting the country south and east of Fort Scott, and the possibility of your falling in with some of the pirates with which the coast upon the Gulph of Mexico has been infested, render it proper that your men should be kept upon the alert and always ready for action in defence of the vessels and cargo. Any hostile movement or outrage towards either will be repelled with a prompt effort of the skill and prowess of your command.[2]

The general promised to send a detachment in small boats down from the fort to assist Muhlenberg's command in getting up the Apalachicola. It was anticipated that the supply boats would reach the Flint ahead of the infantry, which was marching on foot along a new road through unmapped wilderness, but the foot soldiers would wind up with the easier journey. The supply vessels would not reach Fort Scott until more than one month after the arrival of the infantry.

In one final precaution before leaving the Alabama River, General Gaines requested that his counterpart, the commander of the Western Division of the 8th Military District (Gaines commanded the Eastern Division), assist in the operation by keeping a watchful eye on the border of Florida from the mouth of the Perdido River to the Conecuh. While the general did not consider the risk to be great, he nevertheless asked for the support in the event that the absence of the First Brigade from the vicinity might encourage Red Stick parties from the area around Pensacola to launch raids across the border into Alabama.[3]

Gaines then took up his line of march for the Chattahoochee, leading a small bodyguard ahead of the advancing columns of soldiers so he could reach Fort Gaines and then Fort Scott as quickly as possible. The advance was covered in the newspapers of the day:

MILLEDGEVILLE, (Geo.), Nov. 4.

We have no recent information from General Gaines. The last accounts of his contemplated expedition against the hostile Florida Indians, left his whole regular force on their march for Fort Scott, which is in the vicinity of the unfriendly savages. It is not improbable but his approach will inspire such dread as to cause them to sue for peace, and thus prevent the effusion of blood.[4]

It took about two weeks for General Gaines to reach the fort that bore his name, but he reported his arrival there to Major General Andrew Jackson on November 9, 1817. The soldiers were still slowly marching across Alabama in his wake while ahead from Fort Scott, Major Twiggs was warning that trouble was looming on the horizon:

From Major Twiggs I learn that he has received information, upon which he places reliance, that the Indians have recently had a meeting at the Mickasuky town of near 2,700 warriors, when it was determined they would attack us as soon as we should cross the Flint river. Although I put little faith in these threats, and believe their numbers to be overrated, yet I deem it proper, keeping an eye to the safe side, to be provided with additional force; and have, therefore, desired the Governor of Georgia to send me the regiment of infantry and squadron of cavalry held in readiness for that purpose; for, in a war with savages, I think little should be hazarded, as every little advantage which we suffer them to acquire tends to add, in an extraordinary degree, to their strength and confidence.[5]

While the general continued to Fort Scott, his message seeking militia reinforcements traveled across Georgia to Milledgeville. Governor William Rabun would comply with the request from General Gaines and order state militia forces to begin assembling for operations in support of the regular army troops. Additional strength was sought from Charleston, where an officer arrived during the first week of November to retrieve the four pieces of field artillery left behind by Captain Donoho's company:

INDIAN NEWS. – We learn by an officer of the army who arrived in town on Sunday last, direct from Fort Hawkins, that hostilities are expected immediately to break out between our troops and the Seminole Indians. General Gaines, with his force, took up the line of march from Fort Montgomery, on the Alabama, to Fort Scott, on the Flint River, about the 27th ult. Where he was to be joined by 600 Creek Indian warriors, who would make his army, including regulars, militia and Indians, amount to about 2,500 men. The Seminoles are said to have in the field 1,500 warriors. Gen. Jackson and suite, it was expected, would join Gen. Gaines at Fort Scott.[6]

The reference to Creek Indian warriors was to the forces of William McIntosh, which were being called out to support the general movement to Fort Scott. They would not advance, however, until early the following year.

General Gaines reached Fort Scott in safety the following week but was surprised to find that the supply boats under Major Muhlenberg had not yet arrived. The First Brigade by then had reached Fort Gaines and was already on the march for the Flint River. The general immediately dispatched a courier down the Apalachicola to Muhlenberg, carrying a message that urged him to push on for Fort Scott as quickly as possible:

The waters having risen sufficiently high to enable you to ascend the river with all the vessels, I wish you to do so, though it should take longer than I had anticipated. You can avail yourself of the aid of Lt. Scott's detachment to expedite your movement hither. Keep your vessels near to each other, and should you meet with any unsuperable obstacle endeavor to apprise me thereof and you shall have additional relief.[7]

The delay had been caused by low water in the Apalachicola. The two ocean-going supply vessels were too deep in draft to ascend the river until the water had risen. Consequently, Muhlenberg was forced to remain at anchor in Apalachicola Bay until the river rose sufficiently to become navigable.

The message to Major Muhlenberg was carried by Lieutenant Richard W. Scott. Likely because he had earlier brought provisions up the Apalachicola by boat, Scott was now chosen to go down with a detachment of around 40 men to assist the supply vessels in getting up the river. The soldiers who went down with him were all from the 7th Infantry, most of them from Major Twiggs' company, to which Scott also was assigned. They went down in a large boat that was supplied with oars so it could be navigated either up or down the river by manpower.

The main bodies of the 4th and 7th Infantry regiments reached Fort Scott on November 19, 1817. In the days preceding, General Gaines had asked Neamathla to come to the fort for a discussion. The chief declined, however, indicating to the general's messenger that he had already said to Major Twiggs all that he had to say. With his troops now on hand, Gaines determined to overcome the chief's recalcitrance by using force. Orders to that effect were immediately prepared in writing for Major Twiggs:

The hostile character & Conduct of the Indians of the Fowl Town, settled within our limits, rendering it absolutely necessary that they should be removed, you will proceed to the town with the detachment assigned you, and remove them. You will arrest and bring the chiefs and warriors to this place, but should they oppose you, or attempt to escape, you will in that event treat them as enemies. Your men are to be strictly prohibited, in any event, from firing upon, or otherwise injuring, women and children.

You will return to this place with your command as soon as practicable.

Should you receive satisfactory information that any considerable number of the neighboring Indians have joined those of Fowl Town, you will immediately return to this place without making any further attempt to execute first the above written orders.[8]

To carry out his mission, Twiggs was assigned a force of 250 men. While it has long been known that the force included men from both the 4th and 7th Infantries, letters in the National Archives reveal that a few men from Captain Donoho's Company, 4th U.S. Artillery, also accompanied the expedition. Since their field guns had not yet reached Fort Scott, they marched with the rest of the infantry.[9]

Leaving Fort Scott on the evening of the 20th, Twiggs marched his command up the west bank of the Flint to the crossing where Burges's Town had stood until about a decade earlier. The ruins of its cabins and other structures could still be seen in 1817, covering the top of the bluff in the area of today's Oak City Cemetery in Bainbridge. The soldiers crossed the river under cover of darkness and turned south on the old Pensacola to St. Augustine Road, advancing quietly to Fowltown which stood near today's Bainbridge Country Club on the fringe of Fowltown Swamp.[10]

Neamathla had no reason to expect that an attack was imminent and had not positioned warriors to act as sentries on the road leading from the Burges's Town crossing to his village. This allowed Twiggs and his troops to approach Fowltown in the night without detection. They found the town situated on a peninsula that was nearly surrounded by the wetlands of Fowltown Swamp. The site today looks considerably different than it did in 1817, due to the partial clearing and draining of the swamp over the intervening years.

Hoping to surround the town and trap its inhabitants before they learned of his presence and had a chance to escape, Twiggs ordered the companies of Major Montgomery and Captain Burch to begin moving to his right. The companies of Captains Allison and Bee were ordered to begin an encirclement of the town to the left. Twiggs and his company remained in position just north of the town to form the anchor upon which both of these encircling wings rested.[11]

It was in the darkest hours before the dawn on the morning of November 21, 1817, and the operation was going as hoped when the warriors of Fowltown suddenly learned that U.S. troops were moving into position around their town:

...Having marched all the night of the 20th I reached the town before day light on the morning of the 21st & posted the troops in order of battle intending silently to surround it & without blood shed bring to you the chiefs & warriors, but they fled from the companies of Majr. Montgomery & Capt. Burch on my right, & fired upon my left under Capts. Allison & Bee. When they were fired on in return, discovering my superiority of force they fled to a neighboring swamp.[12]

The exchange of fire could not really be termed a battle, as Neamathla and his warriors fired only once, and the troops likewise replied with only a single volley. None of the soldiers were wounded and Indian losses were uncertain. General Gaines reported the chief's losses at four killed and a number wounded, but his source for this information is unclear, Major Twiggs reported, "I had not a man killed or wounded & the Indians but few as they received but one round & fled."[13]

Peter Cook, the clerk of Alexander Arbuthnot, arrived in the vicinity about two weeks later at the head of a war party of Seminole reinforcements and was told that Neamathla's loss in the first encounter with the troops had been only one killed and two wounded.[14]

If the information given Cook was accurate, then the first person killed in the four-decade long Seminole Wars could have been a woman who was shot as she fled before the troops of Major Montgomery and Captain Burch:

It is with deep regret I have to add that a woman was accidentally shot with some warriors in the act of forcing their way through our line, formed for the purpose of arresting their flight. The unfortunate woman had a blanket fastened around her, as many warriors had, which, amidst the smoke in which they were enveloped, rendered it impossible, as I am assured by the officers present, to distinguish her from the warriors.[15]

The soldiers occupied the village and maintained a defensive posture until daybreak when they began to explore the various cabins and other structures. Major Twiggs maintained strict discipline over his men and forbade them from looting the town. He probably knew that the exchange of fire had inaugurated open warfare with the Indians and wished to do nothing more to infuriate them.

It was during this post-encounter search of the village that the officers found among the possessions of the chief himself evidence of his strong affection for the British:

Among the items found in the house of the chief was a British uniform coat, (scarlet,) with a pair of gold epaulettes, and certificate signed by a British captain of marines, "Robert White, in the absence of Colonel Nichols," stating that the chief had always been a true and faithful friend of the British.[16]

Leaving the town and its storehouses of corn and other supplies intact, Twiggs began his return march to Fort Scott. The soldiers took only a few horses and cows as they withdrew. Arriving back at the fort that evening, the soldiers undoubtedly collapsed into their beds. They had marched roughly 30 miles and fought a skirmish in a little over 24 hours.[17]

The unsuccessful attempt to capture Neamathla and his warriors left General Gaines with little doubt that war was now inevitable. In his report of the affair to General Jackson, he noted that friendly chiefs placed the strength of the Red

27

Sticks and Seminoles at around 2,000 warriors, plus perhaps 400 African warriors from the Black Seminole town at the Suwannee. The latter men lived with their families in a separate village but under the overall guidance of Boleck, who had settled with his band of Alachua Seminoles on the west side of the Suwannee in the vicinity of today's community of Old Town.[18]

Based on the reports of his informants, Gaines warned Jackson that the Seminoles and Red Sticks were expecting new support from the British. George Woodbine, who had held the rank of brevet major in Nicolls' command during the War of 1812, was now sailing the Caribbean and Gulf of Mexico in search of adventure and was considering a filibustering expedition to Florida:

> . ...They have been promised, as several Indians inform me, assistance from the British at New Providence. This promise, though made by Woodbine, is relied on by most of the Seminole Indians. I have not a doubt but they will for peace as soon as they find their hopes of British aid to be without foundation.[19]

It is known that Woodbine had visited Tampa Bay not long previous to the eruption of violence at Fowltown, but the promises from him likely were conveyed to the Indians by Robert C. Ambrister. A former lieutenant from Nicolls' command, he had arrived at the mouth of the Suwannee to enlist the participation of the Seminoles and Red Sticks in Woodbine's shadowy plan to seize possession of Florida. Both he and Arbuthnot were on the coast and both were promising additional supplies for the Indians from the Bahamas, although the latter individual seems to have been more interested in trading than adventuring.

After discussing the operation against Fowltown with Major Twiggs and his officers, General Gaines decided to launch a second raid on the town. The plan this time was to examine the surrounding area and bring supplies back to Fort Scott. Numbering 300 men, the force organized for the raid was commanded by Lieutenant Colonel Matthew Arbuckle of the 7th Infantry.

Leaving Fort Scott on the afternoon of November 22, 1817, Arbuckle led his men along the road used two days earlier by Major Twiggs. Marching up the west bank of the Flint to the crossing at Burges's old town, they waded the river and arrived at Fowltown on the morning of the 23rd to find the village abandoned. Neamathla, however, was much more alert this time and knew the soldiers were coming long before they arrived in his town:

...The Lieut. Col. reports, that a party of Indians had placed themselves in a swamp, out of which about 60 warriors approached him, and with a war hoop commenced a brisk fire upon the detachment. They returned the fire in a spirited manner. It continued not more than 15 or 20 minutes before the Indians were silenced, and forced to retire into the swamp, with a loss which Lieutenant Col. Arbuckle estimates at 6 to 8 killed, and a much greater number wounded. We had one man killed, and two wounded.[20]

Once again, the loss sustained by Neamathla's warriors was probably over-estimated by the American officers. Peter Cook reported the Indian loss at 2 killed in this second encounter, which has been remembered as the Battle of Fowltown. It was the first actual battle of the Seminole Wars. According to Cook, the warriors withdrew only because their ammunition had run low, not because they had been driven off by the soldiers.[21]

The U.S. Army also sustained its first casualties of the Seminole Wars at the Battle of Fowltown. Private Aaron Hughes, a young fifer from Marlborough, South Carolina, was killed in the action. One account indicates he was shot down while standing on an Indian house so his fife could be heard above the din of the battle. According to his enlistment records, Hughes stood 5'1" tall and had blue eyes, dark hair and a fair complexion. A farm boy, he had joined the army in June 1814 when he was only 15 years old. He was serving in Major Montgomery's company when he was killed. He was the first U.S. soldier killed in the Seminole wars.[22]

A second account of the Battle of Fowltown appeared in a Milledgeville newspaper on December 9, 1817:

...The detachment consisted of three hundred men, under the command of Col. Arbuckle. They were attacked, about twelve miles from Fort Scott, by a party of Fowltown and Osouchee Indians, supposed to be about 100, and had one man killed and two wounded, one dangerously. The Indian loss was supposed to be eight or ten. They captured some cattle during the fight, which were retaken in the towns lying about eight miles from Fort Scott.[23]

A third somewhat more detailed account was provided by an unidentified officer at Fort Scott, who briefly described the engagement in a letter to his father who lived in Baltimore, Maryland:

...On the 22d, col. Arbuckle crossed Flint river with 300 men, for the purpose of destroying an Indian town about 20 miles off. We arrived in the town about 12 o'clock; next day, at 3, the Indians attacked us, and, after an action of about fifteen minutes, they retreated into a large swamp, which nearly surrounded their town. Their loss cannot be ascertained. Ours, one killed, one severely and three slightly wounded.[24]

The last account places the U.S. loss in the battle at one killed and four wounded, instead of the one killed, and three wounded given in the other reports. It is unclear which number is accurate.

After loading their wagon with corn from the Indian corncribs and rounding up what cattle they could find, the soldiers withdrew back up the Pensacola to St. Augustine Road. Instead of returning to Fort Scott, however, they halted on the bluff that overlooked the Flint River. There, a few hundred yards south of the old Burges's Town site, they erected a blockhouse that Arbuckle named Fort Hughes in honor of the unfortunate fifer. The body of the dead soldier had been brought back from Fowltown and was buried at the fort. While the precise location of the grave is unknown today, it was somewhere in the vicinity of the J.D. Chason Memorial Park in Bainbridge.

It took Arbuckle's men three or four days to build Fort Hughes, which the various vague descriptions describe as a log blockhouse surrounded by a small picket work. The corn and beef seized at Fowltown lasted only long enough to supply the command while it was working on the fort and the colonel and his men returned to Fort Scott on the 27th or 28th with no more provisions than they had carried when they left. A company of around 40 men was left behind under Captain John McIntosh to hold the new fort.

If there had been any hope of avoiding war with the Seminoles and Red Sticks, that hope was now gone. The raids on Fowltown outraged thousands of Indians and large forces of warriors began moving from their towns both to reinforce Neamathla and his men near Fowltown and to block the Apalachicola River so supplies could not reach Fort Scott. They were bent on revenge and their retaliation for the unprovoked attacks would soon take a terrible form.

[1] Maj. Gen. Edmund P. Gaines to Maj. Peter Muhlenberg, October 11, 1817, Office of the Adjutant General, Letters Received, 1805-1821, National Archives.

[2] *Ibid.*

[3] Maj. Gen. Edmund P. Gaines to the Commander of the Western Division, 8th Military District, October 18, 1817, Office of the Adjutant General, Letters Received, 1805-1821, National Archives.

[4] *Providence Patriot*, November 29, 1817, p. 2.

[5] Maj. Gen. Edmund P. Gaines to Maj. Gen. Andrew Jackson, November 9, 1817, *American State Papers*, Indian Affairs, Volume II, p. 160.

[6] Report dated Charleston, S.C., November 10, 1817, from the *Hampshire Gazette*, November 26, 1817, p. 2.

[7] Maj. Gen. Edmund P. Gaines to Maj. Peter Muhlenberg, November 18, 1817, Office of the Adjutant General, Letters Received, 1805-1821, National Archives.

[8] Maj. Gen. Edmund P. Gaines to Maj. David E. Twiggs, November 20, 1817, Office of the Adjutant General, Letters Received, 1805-1821, National Archives.

[9] Lt. Milo Johnson to Maj. Gen. Edmund P. Gaines, November 30, 1817, Office of the Adjutant General, Letters Received, 1805-1821, National Archives.

[10] Cabins of "Old Settlement" of Burges's Town are shown on the Original Land Lot Surveys of Early County, Georgia, 1819-1820, Georgia State Archives.

[11] Maj. David E. Twiggs to Maj. Gen. Edmund P. Gaines, November 21, 1817, Office of the Adjutant General, Letters Received, 1805-1821, National Archives.

[12] *Ibid.*

[13] *Ibid.*

[14] Peter Cook to Miss Elizabeth Carney, January 19, 1818.

[15] Maj. Gen. Edmund P. Gaines to Maj. Gen. Andrew Jackson, November 21, 1817, *American State Papers*, Indian Affairs, Volume II, p. 160.

[16] *Ibid.*

[17] Maj. David E. Twiggs to Maj. Gen. Edmund P. Gaines, November 21, 1817, Office of the Adjutant General, Letters Received, 1805-1821, National Archives.

[18] Maj. Gen. Edmund P. Gaines to Maj. Gen. Andrew Jackson, November 21, 1817, *American State Papers*, Indian Affairs, Volume II, p. 160.

[19] *Ibid.*

[20] Maj. Gen. Edmund P. Gaines to Gov. William Rabun, December 2, 1817, published in the *Independent American*, January 14, 1818, p. 2.

[21] Peter Cook to Miss Elizabeth Carney, January 19, 1818.

[22] Enlistment Records of the U.S. Army, 1798-1815, National Archives.

[23] Milledgeville *Reflector*, December 9, 1817.

[24] Letter from an officer at Fort Scott to his father, December 2, 1817, *Massachusetts Spy*, December 31, 1817, p. 2.

Illustrations

Chapters One - Four

Lt. Col. Edward Nicolls
(Later in Life as a General)

Site of the "Negro Fort"

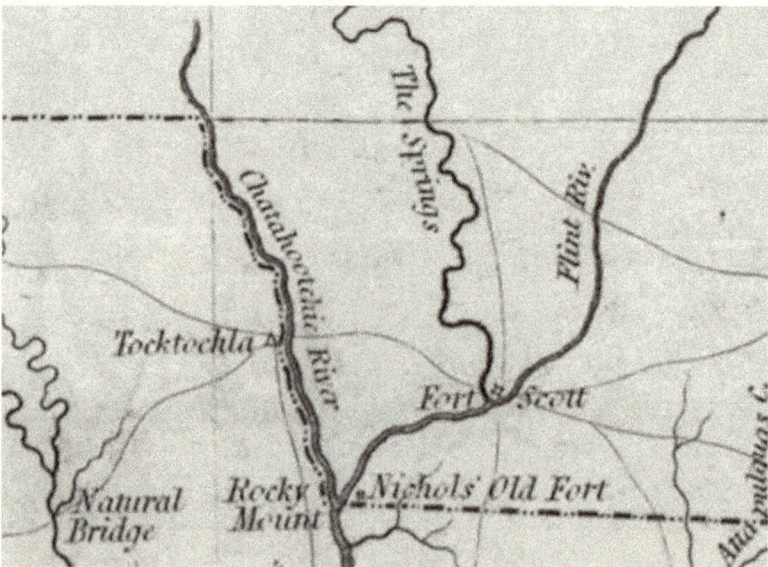

1823 Map of the Confluence of the Chattahoochee & Flint Rivers

Neamathla
Library of Congress

Major David E. Twiggs, as photographed in 1860, 43 years after the Scott Battle.
Library of Congress

Flint River at Bainbridge, Georgia

Original Georgia Land Lot Survey showing Fowltown Swamp

Site of Fowltown

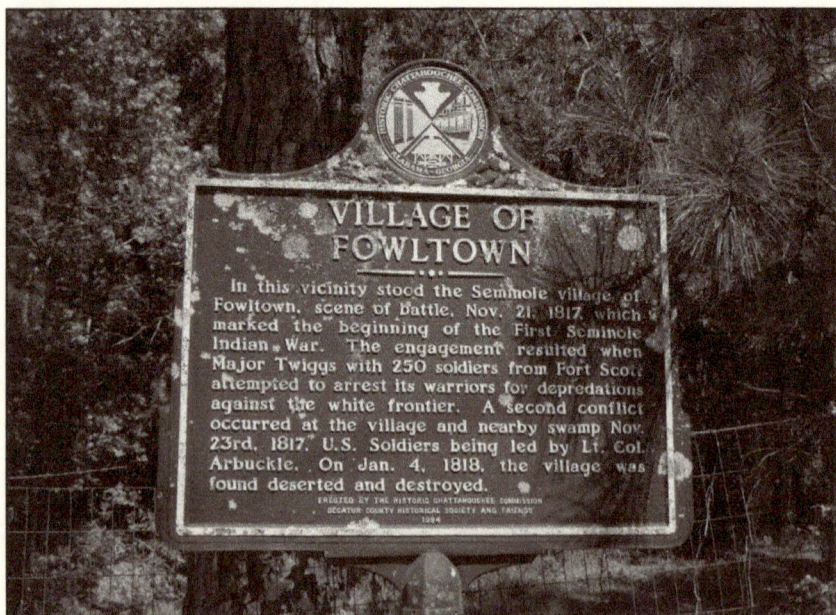

Historical Marker near Site of Fowltown

Lt. Col. Mathew Arbuckle

Site of Fort Hughes in Bainbridge, Georgia

Federal Monument at Site of Fort Hughes

Chapter Five

Prelude to the Attack

MAJOR PETER MUHLENBERG HAD NO WAY OF KNOWING THAT HOSTILITIES had been opened with the Indians. Commanding the supply ships that were trying to make their way up the Apalachicola River, he had been disappointed that General Gaines had not sent more substantial help down the river to him. Lieutenant Scott and his command had reached the flotilla, but he had with him only 40 men and a single boat. Muhlenberg had hoped for more men and more boats

Additional small boats would have allowed the major to lighten the cargoes of his larger vessels. This would reduce the drafts of the sailing ships and make it easier for the sailors and soldiers to get them up the unpredictable river. The boat brought down by Lieutenant Scott, however, could not carry sufficient cargo to make much of a difference. Unaware that fighting had broken out at Fowltown, Muhlenberg ordered the lieutenant's boat to be loaded with regimental clothing, which he knew was needed at Fort Scott. He then replaced twenty of Scott's able bodied men with a number of sick and unarmed soldiers, seven women and four children. The lieutenant was ordered to return back upriver to the fort with a written request from Muhlenberg for more help.

Lieutenant Scott and his party encountered no difficulty on the first part of their journey back upriver. There seemed to be more warriors than normal along the banks of the Apalachicola, but they did not fire on or otherwise interfere with the passage of the boat. The men and women on board still did not know about the fighting at Fowltown, but that changed when they reached Spanish Bluff where the Forbes & Company traders Edmund Doyle and William Hambly made their homes.

The two men had lived among the Indians on the Apalachicola since long before the War of 1812. Doyle, in fact, had suffered great financial loss when the British occupied Prospect Bluff on the lower river. Hambly, on the other hand, had served as a lieutenant in the Colonial Marines and was left in charge of the arms and ammunition left behind at the "Negro Fort" when Lieutenant Colonel Nicolls was ordered to withdraw from the river. A practical man, Hambly soon realized that the British would not come back to help restore the lands lost by the Indians in the recent wars and he ended his association with the settlement at Prospect Bluff and cultivated friendship with the Americans just across the border. When the fort was blown up by American forces on July 27, 1816, the Seminoles and Red Sticks accused Hambly of having provided the U.S. warships with the location of the magazine. They vowed revenge against him, but he was still alive and at his Spanish Bluff plantation when the war erupted at Fowltown.

Major Muhlenberg had ordered Scott to stop there on his way back upriver to ask Hambly and Doyle to send a courier overland to Fort Scott so that his request for additional men and boats could reach the post with utmost speed. The lieutenant did so but was surprised to learn from the two traders that hostilities had erupted between the Indians and the United States. They told him of the fighting at Fowltown and warned that hundreds of Seminole, Maroon, and Red Stick warriors were converging on the Apalachicola.[1]

The homes of Hambly and Doyle were at present-day Bristol, Florida, directly across the river from the village of John Blunt, an Upper Creek chief who had fled Alabama during the Creek War of 1813-1814. Often called Lafarka in U.S. military reports, even though the various Muskogee and Hitchiti languages spoken by the Creeks did not include the sound of the letter "r," Blunt was among the chiefs who allied themselves with the British. When news came of the signing of the Treaty of Ghent, however, the English departed and like Hambly and Doyle, John Blunt knew they would not be coming back. He made his peace with the Americans while also trying to remain friendly with his old Red Stick friends including Neamathla, Josiah Francis, Homathlimico, Atasi Mico and Peter McQueen.

Blunt's village, the name of which is preserved in today's Calhoun County seat of Blountstown, stood on the high ground on the west side of the Apalachicola opposite the farms of William Hambly and Edmund Doyle. When Scott's vessel arrived, the traders summoned the chief across to meet with the lieutenant and arrange for Major Muhlenberg's letter to be carried safely to Fort Scott.

The three men together emphasized to Lieutenant Scott the danger of his situation and urged him to delay his continued voyage up the river until additional support could be sent down from the fort. The Red Sticks and Seminoles were intent on attacking the boats on the river, they warned, and warriors were arriving on the Apalachicola from as far away as Boleck's Town on the Suwannee. Scott decided to inform General Gaines of his situation, but declined to halt or even delay his voyage up the river:

> *Mr. Hambly informs me that the Indians are assembling at the junction of the river, where they intend to make a stand against those vessels coming up the river; should this be the case, I am not able to make a stand against them. My command does not exceed forty men, and one half sick, and without arms. I leave this immediately.*[2]

The lieutenant clearly understood the danger that awaited him on the upper Apalachicola River, but why he did not delay his voyage to give time for his letter to reach General Gaines is a mystery that has troubled the minds of military officers and researchers down through the ages. Hambly, Doyle and Blunt certainly warned him that his small force could not hope to stand against the army of warriors assembling on the river.

It is possible that Scott underestimated the time it would take for his men to row their way up to the confluence of the Flint and Chattahoochee Rivers, expecting that his message would have reached Gaines long before his boat reached the point of danger. It also is possible that it took the courier longer to reach the fort than normal because he had to avoid the war parties that were advancing on the river. Two days was a longer time than normal for an Indian courier to make it from Spanish Bluff to Fort Scott, but it took that long to deliver the Muhlenberg's and Scott's letters to the fort. Perhaps everyone simply underestimated the time it would take the messages to reach Fort Scott.

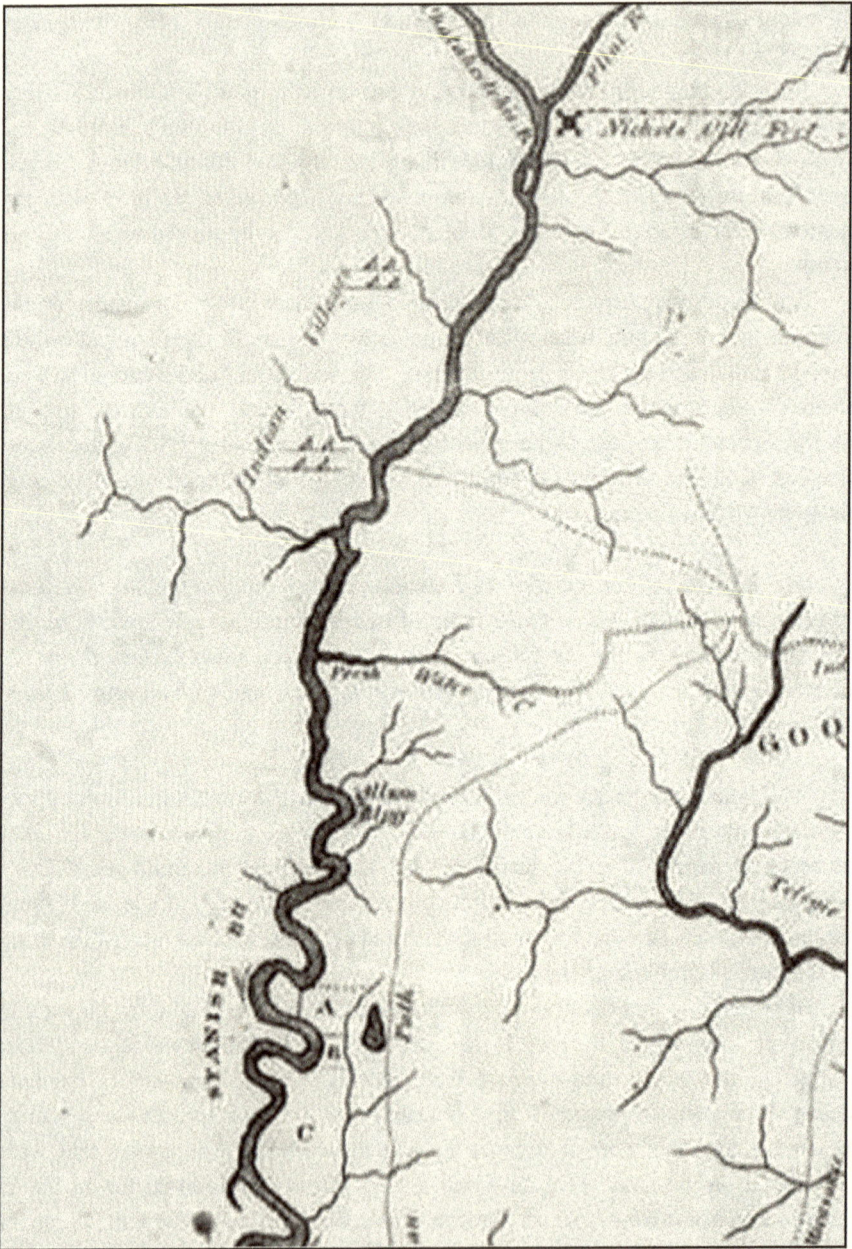

1821 map of the Apalachicola River, showing plantations of Hambly and Doyle (Marked A & B) and the confluence of the Chattahoochee and Flint Rivers.

The other possibility is that the lieutenant underestimated the severity of the threat that awaited him near the confluence, but as so many other military officers have done through the centuries, overestimated his ability to deal with it. The regular troops of the U.S. Army had not always fared well in combat with the Shawnee and other tribes north of the Ohio, as St. Clair's bloody defeat demonstrated. On the Southern frontier, however, it was rare for Native American forces to fight effectively against the forces of the regular army. The 39th U.S. Infantry, for example, had stormed the carefully constructed fortifications of the Red Sticks at the Battle of Horseshoe Bend, killing so many of the defending warriors that the Tallapoosa River ran red with blood. Perhaps Scott believed, as did many other officers, that Creek and Seminole warriors could not overwhelm a regular army command.

It is also possible that he thought the fact that his party was traveling by boat was in itself sufficient protection from an Indian attack. He might encounter annoying fire from the banks, but unless the warriors could somehow get to his vessel across the open water of the Apalachicola, there was no way they could do much damage to his command.

Whatever his reasoning, Lieutenant Richard W. Scott made a deadly mistake on November 28, 1817. He continued up the Apalachicola River from Spanish Bluff, all the while knowing that Red Stick, Seminole and African warriors were waiting for him. His letter to General Gaines did not even request that additional troops be sent down to help secure his passage to the fort, even though he noted from Spanish Bluff that there was "a report here that the Indians have beaten the white people."[3]

The general himself later pondered this same question: Why had Lieutenant Scott continued on after learning the magnitude of the danger that lay ahead?

...It is due to Major Muhlenberg, to observe, that at the time he detached the boat, I have reason to believe he was not apprised of any recent acts of hostility having taken place in this quarter. It appears, however by a letter from Lieutenant Scott, received about the hour in which he was attacked, that he had been warned of the danger which awaited him. I must, therefore, conclude that he felt it to be his duty to proceed. Whether he had received from Major Muhlenberg a positive order to this effect, I have not yet learned.[4]

Whether Gaines ever learned a satisfactory answer to the riddle is not known. Since neither Scott nor any of his key men survived the coming attack and since the known letters of Major Muhlenberg shed no light on the mystery, there is simply no way of knowing.

Leaving the plantations of Hambly and Doyle, the men of Lieutenant Scott's party continued their slow but steady progress up the Apalachicola River. The river was running high so it must have been a laborious row for the soldiers. Very little is known about the construction of the vessel other than that it was made of wood and was large enough to carry 50 people plus a cargo of supplies. When the Indian courier reached Fort Scott on November 30th, General Gaines ordered Captain J.J. Clinch to go to the lieutenant's aide in two covered boats. The captain's written instructions specified that he was to "deliver to him a cover for his boat and give him such assistance as in your judgment shall be necessary to secure his party and expedite his movement to this place." This indicates that Scott's vessel was an open boat, although it is worth noting that one of the survivors of the attack later told Thomas Woodward that it did have a small cabin of sorts.[5]

Whether the boat was equipped with a mast and sail in addition to its long oars is not known. It had no mounted cannon that could be used in its defense, although a survivor of Scott's party named Gray mentioned to Thomas Woodward that a small dismounted swivel cannon was being carried to Fort Scott in the boat's cabin.[6]

The only real defenses on board the boat, in fact, were the 20 muskets carried by Scott's able-bodied crewmen. The rest of his party was made up of women, children and unarmed soldiers who were so sick they would be of little use in defending the vessel against attack.

As the boat slowly continued its journey upstream, the nervousness of the men aboard must have increased substantially. Survivors of the attack later told General Gaines that they had seen hundreds of warriors along the banks of the river between Spanish Bluff and the place where they were attacked. "The wounded men who made their escape concur in the opinion," he wrote in a letter to Governor William Rabun of Georgia, "that they had seen upwards of five hundred hostile Indian warriors at different places, below the point of attack."[7]

The odds against Lieutenant Scott's little party were growing by the hour, but still he continued his journey up the Apalachicola River. By the morning of November 30, 1817, he and his men were just a few miles below the confluence of the Flint and Chattahoochee Rivers. Only a handful of those on board the boat would live to see another day.

[1] Lt. Richard W. Scott to Maj. Gen. Edmund P. Gaines, November 28, 1817, *American State Papers*, Indian Affairs, Volume II, p. 688.

[2] Lt. Richard W. Scott to Maj. Gen. Edmund P. Gaines, November 28, 1817, *American State Papers*, Indian Affairs, Volume II, p. 688.

[3] *Ibid.*

[4] Maj. Gen. Edmund P. Gaines to Gov. William Rabun, December 2, 1817, published in the *Independent American*, January 14, 1818, p. 2.

[5] Maj. Gen. Edmund P. Gaines to Captain J.J. Clinch, November 30, 1817, Adjutant Generals Office, Letters Received, 1805-1821, National Archives; Gen. Thomas Woodward to Col. John Banks, June 16, 1858, in Thomas Woodward, *Woodward's Reminiscenses of the Creek, or Muscogee Indians, contained in letters to friends in Georgia and Alabama*, 1859.

[6] Gen. Thomas Woodward to Col. John Banks, June 16, 1858.

[7] Maj. Gen. Edmund P. Gaines to Gov. William Rabun, December 2, 1817, published in the *Independent American*, January 14, 1818, p. 2.

Chapter Six

The Scott Battle of 1817

THE BLOODIEST DAY OF THE FIRST SEMINOLE WAR dawned not much different than other recent days on Florida's Apalachicola River. Mount Tambora still exerted its influence on the weather of the world and temperatures along the border of Spanish Florida were falling to levels lower than normal.

Aboard the open vessel commanded by Lieutenant Richard W. Scott of the 7th U.S. Infantry, men, women and children shivered in the early morning mist. Some of the soldiers shivered from the cold, but nearly half of them shook with the fever that had overcome them on their long journey from the Alabama River to the Apalachicola. On shore and hidden in the trees where they could not benefit even from the meager sunlight of the morning, Creek, Seminole, Yuchi (Euchee) and African warriors shivered as well. It takes time for the sun to rise high enough over the bluffs that tower above the east bank of the river for the woods and swamps below to benefit from its warming rays.

Roughly one mile below the original confluence of the Chattahoochee and Flint Rivers, the course of the Apalachicola makes a wide bend. The site of the confluence is now beneath the waters of Lake Seminole just above the Jim Woodruff Dam. One mile below the dam, however, the river still swings around

51

the same bend as it begins its southward flow to the Gulf of Mexico. A panoramic view of the curve of the river can be seen from the dock at Chattahoochee Landing and in 1817, as a boat rounded the bend from the South, it would have been possible for its passengers to see straight up the channel to the point of land formed by the confluence.

As Lieutenant Scott's boat came around the bend on the morning of November 30, 1817, the men and women aboard it likely gave a sigh of relief as they spotted the confluence just two miles ahead. They had just passed the landing places of the adjoining villages of the friendly chiefs Mulatto King and Yellow Hair. Located on the Jackson County side of the river just north of today's Gulf Power Company plant, these villages had stood on the high ground overlooking the river swamps since at least the days of the American Revolution. Both towns remained friendly to the United States throughout the First Seminole War, but so far as is known not a single person from either village was seen by the people of Scott's party as they passed by the landings.[1]

As the boat entered the widest part of the arc of the bend, it was pushed hard by the full force of the water pouring from the river's two main tributaries. The Apalachicola was beginning its winter rise, a fact that made its current even stronger. The vessel was pushed from the center of the river towards the east bank as the men pulled hard on their oars to maneuver it against the current and around the bend. Their forward progress stalled as the current ran hard against the side of the boat and drove it ever closer to the bank. All that could be seen there were the trees and bushes of the swamp and the focus of the lieutenant and his men was devoted almost entirely to the navigation of the large bend so that they did not run aground in the shallows.[2]

The chill of the morning replaced by the heat of the adrenalin running through their veins, hundreds of warriors waited in the thick trees and brush that lined the east bank at the point where the boat would be forced closest to shore. Stripped for battle and painted in their traditional colors and designs, they took careful aim with their rifles and muskets and waited for the signal to open fire.

The chiefs and warriors who lined the bank that morning came from a variety of towns and even spoke a variety of languages. There were Hitchiti speakers from Miccosukee, Fowltown and Attapulgas; Muskogee speakers from the Red Stick bands of Homathlemico, Atasi Mico and others; and even the Yuchi-speaking followers of Yuchi Billy, who had come down from their new village in the Old Fields on the high ground to the west of the confluence where the West Bank Overlook stands today. Then there were the African warriors or Black

Seminoles. Some of these men spoke English, some spoke Spanish, a few from the Mobile area spoke French and others had arrived so recently from Africa that they still spoke the tongues and dialects of their native continent.[3]

The overall command of the Indian force was attributed at the time to Homathlemico. A Red Stick chief from Alabama, he had fled his home region following Andrew Jackson's victory at Horseshoe Bend. Escaping south into Spanish Florida along with Peter McQueen, Josiah Francis and others, he had managed to keep most of his warriors together and had allied himself with Lieutenant Colonel Nicolls and the British during the remaining months of the War of 1812. The chief has often been confused for the Atasi Mico, another Red Stick leader, but they were separate individuals.[4]

While Homathlemico probably did lead the attack, the force that assembled to attack Lieutenant Scott's boat did not operate with the same degree of command and control seen in a regiment or brigade of the U.S. Army. The warriors from each town or band fought grouped together and under the leadership of their own war chief. The warriors of each of these groups understood the strategy and tactics by which the battle was to be fought, but they functioned more as independent war parties fighting together to achieve a common objective than they did as individual parts of a cohesive organization. On November 30, 1817, however, the organization of the Indian force did not matter as much as its size. Lieutenant Scott had only 20 able-bodied men, while the total strength of Homathlemico's command cannot even be accurately estimated. He must have had at least 500 warriors at his disposal, probably many more. This gave the attackers a numerical superiority of more than 25 to 1 over Scott's command. The difference in firepower was devastating.

Lieutenant Scott and his men were focused almost entirely on getting their boat around the bend and into the straight channel that would take them up to the confluence when the east bank of the Apalachicola River suddenly erupted with a solid wall of flame:

[The survivors] report that the strength of the current, at the point of the attack, had obliged the lieutenant to keep his boat near the shore; that the Indians had formed along the bank of the river, and were not discovered until their fire commenced; in the first volley of which Lieutenant Scott and his most valuable men fell.[5]

The explosion of gunfire from the trees and bushes along the bank all but annihilated the able-bodied portion of Scott's command. The lieutenant and most of his armed men went down without ever firing a shot. The boat now floated on the current and in minutes was pushed aground in the shallows. The various war cries of the Red Stick Creek, Seminole, Yuchi and African warriors rose above the scene, drowning out the terrified screams of the women and children of Lieutenant Scott's party.

Among the soldiers on the boat that day was a man identified only by his last name, Gray. Badly wounded in the first volley, he was still at Fort Scott when Major General Andrew Jackson arrived there in March 1818 at the head of a brigade of Georgia militiamen. In the campfires of the army camps, Gray described the speed and ferocity with which the attack took place:

...As those on board were hooking and jamming (as the boatmen called it) near the bank, and opposite a thick canebrake, the Indians fired on them, killing and wounding most of those on board at the first fire. Those not disabled from the first fire of the Indians made the best fight they could, but all on board were killed except Mrs. Stuart and two soldiers Gray, and another man whose name I have forgot, if I ever knew it; they were both shot, but made their escape by swimming to the opposite shore.[6]

Gray's account of the battle was preserved by Major Thomas Woodward, a noted frontiersman and Georgia militia officer who served with William McIntosh's Creek Brigade during Jackson's Florida campaign. While his memories of the soldier's story were not perfect when he wrote them down forty years later in a letter to John Banks, a Georgia militia soldier who also served in the invasion, Woodward's account remains the only known detailed description of the battle as originally provided by a participant.

Either six or seven soldiers actually survived the battle, not just the two remembered by Woodward, but otherwise the details of his account are consistent with the reports of the fight sent by General Gaines to the Secretary of War, General Jackson and Governor Rabun of Georgia. The "Mrs. Stuart" mentioned by Woodward was Mrs. Elizabeth Stewart, the wife of a soldier from the First Brigade. Her husband was not part of Scott's party, having taken part in the land march to Fort Scott. His wife and six other women (also wives of soldiers) were on board the vessel, along with four small children. Of these eleven civilians, only Mrs. Stewart survived:

...Lieut. Scott and his Party...were fired on by a party of Indians about two miles from the mouth of the river, and without being able to make any defence fell into their hands, except seven, six of whom came in the succeeding day (five of them wounded). The seventh I understand is at this time with some friendly Indians. The women and children were all killed at that time or since murdered except one who not being wounded is at this time a prisoner with them.[7]

According to Woodward's memories of Gray's account, the warriors waded into the water and stormed the boat as it ran aground in the shallows on the east side of the river. Because Lieutenant Scott and most of his able-bodied men had gone down in the first volley, there was little the sick soldiers, women and children on the vessel could do to defend themselves. In fact, Gray related that the only real resistance put up by the men of Scott's command came in the form of a remarkable last stand by Sergeant Frederick McIntosh from Twiggs' Company, 7th U.S. Infantry:

...When he found all on the boat were lost, and nothing more could be done, he went into a little kind of cabin that the Lieutenant had occupied as his quarters, in which was a swivel or small cannon; loaded it, took it on deck, and resting the swivel on one arm ranged it as well as he could, and (the Indians by this time were boarding the boat) with a fire-brand, he set off the swivel, which cleared the boat for a few minutes of Indians. At the firing of the swivel he was thrown overboard and drowned, and this clearing of the Indians from the boat for a short time gave Gray a chance to escape.[8]

According to Woodward, Sergeant McIntosh was a well-known figure on the Florida frontier. A member of the force that invaded Florida in support of the Patriot Revolt of 1812, McIntosh was described by Woodward as a Scotchman who had served in Colonel Thomas A. Smith's unsuccessful investment of St. Augustine in 1812. Popular with both officers and the enlisted men, McIntosh was said to be a cousin of William McIntosh, the Coweta chief who served under Jackson in both the Creek War of 1813-1814 and the First Seminole War. Woodward remembered that, "Sergeant McIntosh was a man of giant size, and perhaps more bodily strength than any man I have known in our service."[9]

Woodward, as should be expected, was both accurate and inaccurate in his memories of Sergeant McIntosh as recorded 40 years after the man's death. According to his actual enlistment records, McIntosh was enlisted in the U.S. Rifles by Colonel Thomas Smith between February and April of 1813 for the duration of the War of 1812. Woodward was correct in his memory that McIntosh had been born in Scotland and that he had served in support of the Patriot Revolt in East Florida. His enlistment record confirms that he was from Scotland and that he had enlisted at Camp New Hope in East Florida.[10]

On the other hand, the sergeant was not a large and powerful man as remembered long after the fact by Thomas Woodward. His enlistment record indicates that he was 5'10" tall with blue eyes, fair hair and fair complexion when he enlisted at the age of 27 in 1813. Discharged at the end of the War of 1812, the sergeant left the service for a time but then reenlisted on March 5, 1817, in Twiggs' Company, 7th Infantry, as a sergeant. He came back into the army as a substitute for a soldier with the curious name of Young Blood.[11]

The sergeant's act of heroism in taking up the small cannon in his bare hands and firing it at the warriors as they swarmed over the bulkheads of the boat cost him his life, but created an opportunity for Gray and the other five male survivors to escape. The soldiers able to do so, all but one of whom were wounded, went over the sides of the vessel and swam away under the water. They swam across the current of the Apalachicola and pulled themselves from the river on the Jackson County shore. Others might also have attempted to escape in this way, but if they were severely wounded, the river could have certainly claimed their lives before they reached the opposite bank.

The survivors appear to have been rescued by the inhabitants of the villages of Mulatto King and Yellow Hair on the Jackson County side of the river. As was noted earlier, the chiefs and most of the people of these towns tried to remain at peace with the United States during the war. In notifying Major Muhlenberg of the disastrous fate of Lieutenant Scott's command on December 2, 1817, Major Clinton Wright noted that one of the surviving men was then "with some friendly Indians."[12]

Things did not go well for the wounded men, women and children trapped on the boat. The fate of the most helpless of these was particularly gruesome. Peter Cook, the store clerk of Alexander Arbuthnot, heard accounts of the attack first hand from some of the Seminole or Red Stick participants who described how the children aboard Scott's vessel were put to death. He repeated these descriptions in a letter sent from the Suwannee River to Miss Elizabeth Carney in the Bahamas just six weeks after the Scott Battle:

There was a boat that was taken by the Indians, that had in it thirty men, seven women, four small children. There were six of the men got clear, and one woman saved, and all the rest of them got killed. The children were took by the leg, and their brains dashed out against the boat.[13]

The men and women found wounded in the boat fared no better. Elizabeth Stewart later told officers on Jackson's staff that the attacking warriors severed the breasts of the six slain women before scalping and otherwise mutilating their remains. Lt. Scott, she reported, was found badly wounded but still alive by Homathlemico and his men:

...Lieutenant Scott (as described by the woman prisoner) was tortured in every conceivable manner. Lightwood slivers were inserted into his body and set on fire, and in this way he was kept under torture for the whole day. Lieutenant Scott repeatedly begged and importuned the woman that escaped the slaughter to take a tomahawk and end his pain. But 'No,' said she, 'I would as soon kill myself.'[14]

Killed with hatchets and clubs, they were scalped and their bodies otherwise mutilated. Scalps recognized by the hair as having belonged to the men, women and children of Scott's party were later found adorning a wooden pole in the primary Seminole villages near Lake Miccosukee.

Of the women, later reports indicate that at least two survived the initial bloodshed. One of these was Elizabeth Stewart:

...Mrs. Stuart was taken almost lifeless as well as senseless, and was a captive until the day I carried her to your camp. After taking her from the boat, they (the Indians) differed among themselves as to whose slave or servant she should be. An Indian by the name of Yellow Hair said he had many years before been sick at or near St. Mary's, and that he felt it a duty to take the woman and treat her kindly, as he was treated so by a white woman when he was among the whites. The matter was left to an old Indian by the name of Bear Head, who decided in favor of Yellow Hair.[15]

The identity of the other woman carried away from the scene of the attack was not recorded and little is known of what happened to her other than that she became too exhausted to keep up with the warriors as they withdrew from the river. She was killed.

The veracity of Woodward's account of the captivity of Mrs. Stewart is difficult to assess. The Yellow Hair mentioned by him could not have been the chief of Yellow Hair's town near the battle scene or she would have been taken back to Fort Scott. The statement that the Yellow Hair responsible for saving the woman did so to repay a kindness paid to him many years before by a woman on the St. Mary's indicates at least circumstantially that he was a Seminole. It is known for a fact, however, that she was with Peter McQueen's band of Red Sticks when she was freed by Andrew Jackson's army following the Battle of Econfina during the spring of 1818.

The exact number of people killed and wounded in the attack on Scott's party is difficult to determine with precision. The military reports and private letters from officers at Fort Scott dating from the days and weeks after the attacks give various estimates. General Gaines himself reported that Scott's vessel was carrying the lieutenant, 40 soldiers and seven women. Of this number, he reported, six men and one woman survived. The general's estimate, then, would place total losses in the attack at 34 or 35 men and six women killed, four men wounded and one woman – Mrs. Stewart – captured.[16]

This tally was generally supported by a statement that Lieutenant Colonel Matthew Arbuckle added to the November 1817 monthly report for the 7th U.S. Infantry:

...Since making out the enclosed report I have recd. The unpleasant news of Lieut. Scott and thirty three or four men, being killed by the Indians on the 30th ultimo; this took place on the Appalachicola River (as the party was ascending in a boat) a short distance below the junction of the Flint and Chattahochie Rivers.[17]

Peter Cook, writing from information provided to him by Indian warriors involved in the attack, estimated the U.S. loss at 24 men, 6 women and 4 children killed. His account agrees with that of General Gaines in its statement that only seven people – six men and one woman – survived the battle. Cook's account provided the lowest number of fatal casualties of the known accounts from the two months after the battle.[18]

The highest estimate appeared in the Savannah *Republican* on December 17, 1817. That account reported that Scott, 44 of his men, 10 women and three children were killed, numbers that combined for a total loss of 58 men, women and children. The paper did not provide information on the source of its data.[19]

A name by name analysis of the enlistment records of the U.S. Army has produced a list of 28 soldiers, including Lieutenant Scott, killed in action, plus one additional man listed as missing and presumed dead. Adding to this number the women and children reported killed and the soldiers reported wounded, a final casualty list can be assembled of 28 men, 6 women and four children killed, 4-5 men wounded, one woman captured and one man reported missing but presumed dead. Either one or two men escaped the attack unharmed. If these numbers are totally inclusive, then the total loss for Scott's party was 38 killed, 4-5 wounded, one missing in action and one captured, or a combined total of 44-45 known killed, wounded, missing and captured.[20]

Other soldiers may also have been killed in the action and the reports of General Gaines and Lieutenant Colonel Arbuckle certainly add support to such a possibility. Both officers placed the total loss in the attack on Scott's party at around 33 or 34 men killed. At the time they wrote their reports, however, they had not been able to communicate with Major Muhlenberg to know how many men from Scott's original detachment had been left behind with that officer on the lower Apalachicola. If the wooden boat assigned to Lieutenant Scott was designed to carry around 40 men, and this is a point worth considering, then Muhlenberg would not have sent it back up to Fort Scott with more people aboard than the vessel was designed to carry.

When the boat left the fort, it was carrying 40 men and no cargo. When the major ordered it back up to the Flint River, he had the clothing stores of the 4th U.S. Infantry placed aboard. When they wrote their reports, Gaines and Arbuckle did not know that part of the vessel's capacity was being used for that purpose. If the boat was designed to carry 40 soldiers, then with part of its burden being used to carry cargo, the smaller estimates of total losses seem more realistic. Peter Cook cited warriors who participated in the attack in placing the total number of people on board Lieutenant Scott's vessel as 30 men, 7 women and 4 children. His number is almost identical to the one that results from a study of the post returns, enlistment records and official reports (i.e. 31 men, 7 women and 4 children).[21]

Without additional information on the size of the vessel, the only conclusion that can be made is that the total U.S. loss in killed, wounded and captured was between 38 and 48. The total verifiable loss was 44-45 with a minimum of 38 men, women and children being killed. The U.S. Army would not experience a bloodier day until Major Francis Dade's party was destroyed eighteen years later at the beginning of the Second Seminole War. It is worth noting that Dade was a 1st lieutenant in the 4th U.S. Infantry when the attack on Scott's party took place.

He was absent on recruiting duty and was not at Fort Scott when news of the disaster was received.[22]

---

[1] Capt. Hugh Young, "A Topographical Memoir of East and West Florida," 1818.

[2] Maj. Gen. Edmund P. Gaines to the Secretary of War, December 2, 1817, *American State Papers*, Indian Affairs, Volume II, p. 687.

[3] List of participating groups assembled from numerous U.S. Army reports, 1817-1818, that identify the village affiliations of members of the attacking force.

[4] Maj. Gen. Andrew Jackson to John C. Calhoun, Secretary of War, April 8, 1818 & April 9, 1818, American State Papers, Military Affairs, Volume 1, pp. 699-700.

[5] Maj. Gen. Edmund P. Gaines to the Secretary of War, December 2, 1817, *American State Papers*, Indian Affairs, Volume II, p. 687.

[6] Gen. Thomas S. Woodward to Col. John Banks, June 16, 1858, *Woodward's Reminisences*.

[7] Maj. Clinton Wright, Assistant Adjutant General, to Maj. Peter Muhlenberg, December 2, 1817, Office of the Adjutant General, Letters Received, 1805-1821, National Archives.

[8] Gen. Thomas S. Woodward to Col. John Banks, June 16, 1858.

[9] *Ibid.*

[10] U.S. Army, Register of Enlistments, 1798-1815, National Archives.

[11] *Ibid.*

[12] Maj. Clinton Wright, Assistant Adjutant General, to Maj. Peter Muhlenberg, December 2, 1817.

[13] Peter B. Cook to Elizabeth A. Carney, January 19, 1818, included in Message of the President of the U. States to Congress, 25th March, 1818, published in the *New York Mercantile Advertiser*, January 6, 1819, p.2.

[14] Surgeon attached to Jackson's staff, quoted in James Parton, *Life of Andrew Jackson*, p. 458.

[15] Gen. Thomas S. Woodward to Col. John Banks, June 16, 1858.

[16] Maj. Gen. Edmund P. Gaines to the Secretary of War, December 2, 1817; Maj. Gen. Edmund P. Gaines to Gov. William Rabun, December 2, 1817.

[17] Lt. Col. Matthew Arbuckle to Brig. Gen. D. Parker and Staff, December 7, 1817, enclosed on the Monthly Report of the 7th U.S. Infantry for November 1817, Adjutant General's Office, Letters Received, 1805-1821.

[18] Peter B. Cook to Elizabeth A. Carney, January 19, 1818.

[19] Savannah *Republican* December 17, 1817.

[20] U.S. Army Enlistment Registers, 1798 -

[21] Maj. Gen. Edmund P. Gaines to the Secretary of War, December 2, 1817; Maj. Gen. Edmund P. Gaines to Gov. William Rabun, December 2, 1817; Peter B. Cook to Elizabeth A. Carney, January 19, 1818; U.S. Army Registers of Enlistment, 1798 - ; Lt. Col. Matthew Arbuckle to Brig. Gen. D. Parker and Staff, December 7, 1817.

[22] U.S. Army Register of Enlistments, 1798-1815, entry for Francis L. Dade.

Illustrations

Chapters Five - Six

19th Century Artist's Rendering of the Attack on Scott's Party

Site of the Scott Battle at Chattahoochee, Florida

Chattahoochee Landing from the Air (Battle Site is at Right)

19th Century Artist's Rendering of the "Attack on Scott"

Actual Scene of the Attack at Chattahoochee, Florida

Maj. Gen. Edmund P. Gaines

Photographs

Site of Fort Scott in Decatur County, Georgia

Site of Fort Scott, Showing Area where Main Gate was Located

Historical Marker for Fort Scott at Hutchinson's Ferry Landing

1820 Land Lot Survey showing Fort Scott

Chapter Seven

Aftermath of the Attack

THE ATTACK ON LIEUTENANT SCOTT'S VESSEL took place about twelve miles downstream from Fort Scott, too far away for the sounds of the rifles of the Indians and muskets of the soldiers to travel. If the blast touched off from the small cannon by Sergeant McIntosh was heard at the fort, none of the officers there made mention of it in their correspondence. The runner sent from the plantations of Hambly and Doyle with Scott's report that he faced growing danger on the upper Apalachciola did not reach Fort Scott until about the hour of the attack on the lieutenant's boat.

The unidentified Indian messenger had traveled overland from Spanish Bluffs to carry the reports of both Muhlenberg and Scott to General Gaines. His journey had taken two days to complete and it was not until November 30, 1817, that the general read the letter penned by Scott two days earlier. Gaines was shocked into immediate action by the lieutenant's statement that he would not be able to resist the force rumored to be waiting for him on the upper Apalachicola River:

Upon the receipt of Lieutenant Scott's letter, I had two boats fitted up with covers of plank, port holes, &c. for defence, detached them under Captain Clinch, with a sub-altern officer and 40 men, with an order to secure the movement of Lieutenant Scott, and then to assist Major Muhlenberg. This detachment embarked late in the evening of the 30th ult. and must have passed the scene of the action (15 miles below this place) at night, and 7 hours after the affair had terminated.[1]

Unaware that Scott's party had been destroyed, General Gaines issued specific orders to Captain J.J. Clinch of the 7th U.S. Infantry, the officer to whom he entrusted command of the relief force. The captain was to move down from the fort with all possible speed and assure the safety of the vessel then believed to still be making its way up the Apalachicola:

You will embark with the party assigned you on board the two covered boats; descend the river until you meet with Lieut. Scott; deliver to him a cover for his boat, and give him such assistance as in your judgment shall be necessary to secure his party and expedite his movement to this place. You will then proceed with the residue of your command down the river until you meet with Major Muhlenberg, report to him, and act under his orders.[2]

The general's instructions were written after it was too late to do anything to save the command of Lieutenant Richard W. Scott, a fact that rendered his final admonition to Captain Clinch even more tragic:

You will in no case put your command in the power of Indians near the shore. Be constantly on the alert – Remember that U.S. troops can never be surprised by Indians without a loss of honor – to say nothing of the loss of strength that might ensue.[3]

The captain and his dangerously small detachment of 40 men left Fort Scott after darkness had fallen on the evening of November 30, 1817. As General Gaines later surmised, the two relief vessels passed the scene of the attack in the night and about seven hours after the battle. If Captain Clinch saw any sign of Lieutenant Scott's boat or of the dead men, women and children who lay in heaps at the site of the encounter, no letter or report from him indicating such has ever been found.

For the entire night of November 30, 1817, no one at Fort Scott knew what had happened to Lieutenant Scott and his unfortunate command. General Gaines undoubtedly hoped that the reinforcements sent down under Captain Clinch were more than sufficient to help the lieutenant complete his passage up to the fort, but on the next day such hopes were replaced by shock:

Yesterday five of his men came in all wounded. They state, that Lieut. Scott was attacked by the Indians just below the forks of the river, and the whole party killed except themselves. This is truly lamentable. I expect we shall have some very warm work before many days. The whole Indian force is supposed to be 2,800.[4]

The final number of survivors that reached Fort Scott is unclear. In his report to the Secretary of War dated December 2, 1817, the general indicated that six had come in, four of whom were wounded. On the same day, however, Major Clinton Wright, the assistant adjutant general assigned to General Gaines, penned a warning to Major Muhlenberg, informing him that six men had come in, five of whom were wounded, while a seventh was with friendly Indians. Whether this seventh individual ever reached the fort is not clear.[5]

The names of only two of the wounded have been identified. The first of these was Private William James from Twiggs' Company, 7th U.S. Infantry. Formerly a corporal who had been reduced in rank, he was reported to have been "wounded in action near Fort Apalachicola." This apparently was the name given by U.S. forces to the outpost built one mile below the confluence of the Chattahoochee and Flint Rivers by British troops during the War of 1812. Constructed in 1814, the fort was evacuated the following spring and so far as is known never was occupied again. [6]

The second known survivor of the battle was the soldier identified only by his last name of Gray. He was mentioned in the account of Thomas Woodward, who reported that Gray was present with Jackson's army during the 1818 Florida Campaign.

It was also known at Fort Scott by the morning of December 2nd that Elizabeth Stewart had survived the attack. In his dispatch to Major Muhlenberg, Major Wright reported, "The women and children were all killed at that time or since murdered except one who not being wounded is at this time a prisoner with them." Information on her location and condition would arrive steadily at the fort over the next several months, providing tantalizing bits of intelligence to the officers there and to her desperate husband, who was among the men stationed at the post. Contrary to some accounts, he was not one of the men killed in the attack.[7]

On the morning of December 2, 1817, the chief John Blunt arrived at Fort Scott with a message from Major Muhlenberg. A reply was immediately sent out, probably via the chief, informing the major of the disaster and warning him to be extremely cautious as he continued up the river with the supply vessels:

...Capt. Clinch must have joined you ere this and having done so, you will make such a disposition of his boats as you may think best, either construct bulwarks with the plank thereof or sink them, but in no case suffer your men to approach the bank of the river which it is presumed you will be compelled to do should you attempt to navigate them up stream.[8]

The two boats sent down on the evening of the 30th under Captain Clinch were apparently propelled by oars only. Without sails to help them navigate upstream against the current, it would have been impossible for the men aboard them to avoid approaching the bank of the river. The two supply vessels under the protection of Major Muhlenberg's command were the sloop *Phoebe Ann* and the schooner *Little Sally*. Designed to navigate the coastal waters of the Gulf of Mexico, both were equipped with sails and could slowly make headway up the Apalachicola River when the winds blew in the right direction. The major was instructed to exercise the utmost caution, but to continue his movement up the river:

You will take the advantage of every wind that will enable you to progress, keeping in the middle of the river. So soon as the militia arrives a movement will be made and arrangements to form a junction with you, either at Spanish Bluff, or below, should you not be able to ascend the river to that point before we reach it. I will again remark your boats must be kept together and in the middle of the river, where with such bulwarks as you will be able to construct it is impossible you can sustain any injury from the species of force you will have to contend with, whatever may be their number.[9]

Even as General Gaines was issuing orders and developing a plan to safely bring Major Muhlenberg and his vessels to Fort Scott, unexpected orders arrived from the Secretary of War directing him to Point Peter on the St. Mary's River. President Monroe had authorized U.S. forces to seize Amelia Island on the Florida coast and Gaines was directed to assume command of the operation. When the orders left Washington, D.C., on November 12, 1817, the administration had no way of knowing that hostilities were about to erupt with the Indians. By the time the dispatch reached Fort Scott on December 2nd, however, the general was engaged in a full-scale war with the Seminoles and Red Sticks.[10]

Despite the critical situation that had descended on the frontier, General Gaines had no choice but to comply with the instructions from the War Department. The dispatch from acting Secretary of War George Graham left him no discretion in the matter and he immediately began taking steps to prepare his command for his departure. On December 4th he issued General Orders to the officers and men at Fort Scott, informing them of his departure and assigning Lieutenant Colonel Matthew Arbuckle of the 7th Infantry to the command of forces in the vicinity:

The whole will be put in readiness for a vigorous attack on the enemy, whose long continued hostility and recent massacre of sick men and helpless women and children, demand and shall receive a full measure of retaliation. The Genl. calculates upon returning in time to participate in the service. In the mean time he tenders to the officers and men of his command his best wishes for their health, military distinction and personal prosperity.[11]

Any plans of marching to Muhlenberg's relief were abandoned and Arbuckle was ordered to do his best to hold his positions and secure the supply vessels under the very trying circumstances. The departure of Gaines from the scene forced the army to move entirely to the defensive at a critical juncture of the conflict.

The Indian alliance, meanwhile, continued to press its response to the opening of the war by the soldiers from Fort Scott. War parties were sent to hover around the fort and they fired on the post as early as three days after the attack on Scott's party. A letter from an unidentified officer to his father reported on December 2, 1817, "Since I have commenced this letter, the Indians have fired upon some women who were washing on the bank."[12]

The soldiers in the fort generally responded to such attacks with barrages of artillery fire. These had the effect of temporarily driving away the Creek and Seminole snipers, but the warriors inevitably returned and continued to harass the garrison until the following spring. The suspension of offensive operations by the army, meanwhile, allowed the chiefs of the war alliance to concentrate their forces for two major attacks. One of these would be launched against Captain John McIntosh and his small garrison of 40 men at Fort Hughes, while the other would attempt to block the Apalachicola and prevent Muhlenberg's supply vessels from reaching Fort Scott.

Even as hundreds of Creek, Seminole and African warriors closed in on these two primary objectives, Edmund Doyle appears to have launched an initiative of his own to restore peace to the frontier before it was too late. Without the knowledge or approval of Colonel Arbuckle, Doyle approached the Atasi Chief, a prominent Red Stick leader, to see if a truce could be negotiated. The chief was receptive to the approach and contact was made with three American leaning chiefs, William and George Perryman and a third individual known only as Johnson but who appears to have been associated with Mulatto King's town. The chiefs offered to approach the officers at Fort Scott to see what could be done.

Colonel Arbuckle agreed to meet with Johnson and the Perrymans and a council was convened on December 8, 1817:

I have understood that Mr. Doyle has had a talk with Ottossee Micko about making peace. I did not ask Mr. Doyle to make this, or any other Talks with the hostile Indians, but I shall be glad if the talk has enduced them to wish for peace, as their Great Father the President of the United States, has always wished for peace with them.[13]

Johnson and the Perrymans told Arbuckle they had been authorized to open negotiations not only on behalf of Atasi Mico, but Cappachimico of Miccosukee as well. If the Miccosukees could be prevented from actively joining the war, the army would dramatically improve its chances of a quick victory. Arbuckle, however, did not mince words with the emissaries, pointing out to them that the Miccosukees had declined to surrender the alleged murderers of Mrs. Garrett and her family. He also noted the hostile expressions and attitudes of Neamathla and his warriors, commenting on what the army considered the unprovoked attack on the soldiers sent to bring the chief back to Fort Scott for a meeting with General Gaines.[14]

Despite these points of contention, however, the colonel told the friendly chiefs that it was not too late for peace:

Should the Mickysoockee & other Chiefs now at war with the United States wish for peace on terms of justice let them send their talks and they shall be heard, but should they hereafter fire on any of our vessels or boats coming up or going down the river, or kill any of our people, their talks will not be heard until they are severely punished.

The army will not move after the hostile Indians for six days unless it is discovered that their talks for peace are not sincere, and if the hostile Indians want peace, I shall expect to hear from them again in that time.[15]

While he gave the war faction a choice between peace and war, Arbuckle carefully reassured the peaceful towns that they would not be attacked, nor would their lands be taken from them. This was an important concession to the friendly chiefs, several of whom were concerned that they might suffer equally with the warring towns. The Treaty of Fort Jackson, which had ended the Creek War of 1813-1814, had penalized friendly and warring Creeks alike by forcing them to agree to massive land cessions:

Should the hostile Indians still want war, nothing is asked of the friendly Indians but to remain Peacibly at their homes and to keep the hostile Indians from among them, and if the army should march by a friendly town, they have nothing to do but hoist a white flag.

Chiefs Johnson & Perrymans, I am well pleased with your conduct as honest men, and with your good intentions in endeavoring to bring about a peace. If you should fail you have only to remain quietly at your homes, where you will be unmolested.

Your Great Father the President of the United States will not take from the friendly chiefs and warriors their land or property but the Indians of the hostile towns will be expected to pay the expenses of the war they have made.[16]

While mention of her fate does not appear in the transcript of the colonel's talk to the friendly chiefs, he apparently did make the safe return of Elizabeth Stewart a priority in his private discussions with them. He noted this in a dispatch sent on the same day to Major Muhlenberg:

A proposition has been made by the Hostile Chiefs through the friendly chiefs Perriman and Johnston for peace, and as an evidence of their desire for peace they say they will not permit their warriors to fire on our vessels ascending the river, that they will send on board the vessels the woman they took from Lieut. Scott's command.

I shall do nothing against those Indians for six days unless I perceive they are not sincere in their propositions.[17]

Arbuckle wisely cautioned Muhlenberg to remain on his guard at all times, expressing optimism that the major would reach Fort Scott with the vessels under his charge within four days. This optimism reflected the colonel's wishful thinking far more than it did the reality of the situation and it would take weeks longer for the supply boats to complete their upriver voyage.

It did not take long for the commanding officer at Fort Scott to receive his reply from the war faction. Just three days after his talk with Johnson and the Perrymans, a major attack was launched on Fort Hughes. Led in part by Peter Cook, a force of several hundred Creek and Seminole warriors surrounded the little fort and opened a fierce assault. Although he commanded only 40 men, Captain John N. McIntosh had so well prepared his defenses that he was able to fend off the attacks without losing a man. After three or four days, the warriors realized they would not be able to take the fort and withdrew. Their casualties also were thought to be light. Colonel Arbuckle immediately sent Captain Donoho with a force of 2 sergeants, 2 corporals and 60 privates to withdraw McIntosh's men safely back to Fort Scott.[18]

There would be no peace with the war towns and Elizabeth Stewart would not be voluntarily returned to her husband and family.

---

[1] Maj. Gen. Edmund P. Gaines to Gov. William Rabun, December 2, 1817, published in the *Independent American*, January 14, 1818, p. 2.

[2] Maj. Gen. Edmund P. Gaines to Capt. J.J. Clinch, November 30, 1817, Adjutant General's Office, Letters Received, 1805-1821.

[3] *Ibid.*

[4] Letter from an officer at Fort Scott to his father in Baltimore, December 2, 1817, published in the *Massachusetts Spy*, December 31, 1817, p. 2.

[5] Maj. Gen. Edmund P. Gaines to the Secretary of War, December 2, 1817; Maj. Clinton Wright to Maj. Peter Muhlenberg, December 2, 1817, Office of the Adjutant General, Letters Received, 1805-1821.

[6] U.S. Army Enlistment Registers.

[7] Maj. Clinton Wright to Maj. Peter Muhlenberg, December 2, 1817.

[8] *Ibid.*

[9] *Ibid.*

[10] George Graham, Acting Secretary of War, to Maj. Gen. Edmund P. Gaines, November 12, 1817, Office of the Adjutant General, Letters Received, 1805-1821.

[11] Maj. Gen. Edmund P. Gaines, General Orders of December 4, 1817, Office of the Adjutant General, Letters Received, 1805-1821.

[12] Letter from an officer at Fort Scott to his father in Baltimore, December 2 1817.

[13] "Talk Delivered on the 10th of Decr. 1817 to three Indian Chiefs," December 10, 1817, transcribed by Captain Alex. Cummings, Office of the Adjutant General, Letters Received, 1805-1821.

[14] *Ibid.*

[15] *Ibid.*

[16] *Ibid.*

[17] Lt. Col. Matthew Arbuckle to Maj. Peter Muhlenberg, December 10, 1817, Office of the Adjutant General, Letters Received, 1805-1821.

[18] Capt. John N. McIntosh to Hon. A. Lacock, February 5, 1819, *American State Papers*, Military Affairs, Volume 1, p, 747; Peter Cook to Miss Elizabeth Carney; "Register for Details for Command from Fort Scott, from the 18th of December, 1817, until the 19th of March, 1818, whilst under the command of Lt. Col. Arbuckle," Office of the Adjutant General, Letters Received, 1805-1821.

Chapter Eight

The Nation Reacts

GENERAL GAINES HAD ALREADY LEFT FORT SCOTT by the time news of the destruction of Lieutenant Scott's party began to spread up the East Coast. Communication lines of the day were slow, and it took days for word of the attack to reach the settled areas of Georgia.

On December 2, 1817, the general dispatched an express rider from Fort Scott to carry his official reports of the battle to Fort Hawkins at present-day Macon so they could be forwarded on to the Secretary of War, Governor William Rabun and others. Other officers took advantage of the opportunity to send hastily written letters of their own and the news began to spread, with the first reports appearing in the Milledgeville, Augusta and Savannah newspapers later that week. By December 17th, the *Savannah Republican* was reporting that the death toll from the attack was worse than first thought:

The Savannah Republican of the 17th ult. States that more lives were lost in the last attack upon Lieutenant Scott, by the Indians, than was at first supposed. It appears that Lieutenant Scott, 44 men, 10 women and three children were killed making in all 58. The clothing for the 4th regiment, under the guard of Scott, was also taken off by the savages[1]

The numbers given by the newspaper are significantly higher than those derived from military reports and records and may have included some duplication or exaggeration. Other reports, for example, placed the number of women aboard Scott's vessel at seven, although the number of children mentioned in the letters of various officers was four and not the three mentioned by the paper.

Most Americans along the eastern seaboard first learned of the attack from such reports, which other newspapers copied from the Georgia publications and republished for readers in their own communities. In the days before the introduction of wire services to the newspaper industry, this was how news was spread from town to town across the entire country. In the case of the Scott Battle, the spread of the story can be traced from paper to paper as it moved its way up the coast to Washington, D.C., Baltimore, New York and beyond. By December 31st, for example, it had reached New England where it appeared in the *Massachusetts Spy* and other newspapers.[2]

Just as it would today, the loss of so many U.S. soldiers and civilians – especially children – resulted first in shock and then outrage among the American people. Calls for vengeance erupted and frustration ensued over the lack of additional news from the far away frontier of Southwest Georgia.

The official reports of the battle also made a predictably slow movement from Fort Scott to Washington. General Gaines penned his first dispatches about the attack on Lieutenant Scott's command on December 2, 1817, sending virtually identical letters to Governor Rabun and the War Department. Two days later, upon learning he had been ordered to the St. Mary's, he followed his first report with a somewhat philosophic dispatch to the Secretary of War:

I would much more willingly devote my time and humble faculties in the delightful occupation of bringing over savage man to the walks of civil life, where this is practicable, without force, than to contribute to the destruction of any one of the human race. But every effort in the work of civilization, to be effectual, must accord with the immutable principles of justice. The savage must be taught and compelled to do that which is right, and to abstain from doing that which is wrong. The poisonous cup of barbarism cannot be taken from the lips of the savage by the mild voice of reason alone; the strong mandate of justice must be resorted to, and enforced.[3]

It was a powerful statement from Gaines, who over the previous year on several occasions had intervened to maintain calm between the Creek Indians and the settlers on the Alabama frontier. A tough, battle-hardened veteran, he often did attempt to avoid hostility with the Indians, although his attempt to bring Neamathla to Fort Scott had in fact instigated the current war.

By November 9th, ten days after the attack on Scott's vessel, news of the incident still had not reached Washington. The War Department learned that day of the initial encounter at Fowltown and responded by expressing hope that future violence could be avoided:

Your letter bearing date the 21st ultimo, advising of the arrival of the first brigade at Fort Scott on the 19th ultimo, and of the subsequent affair with the Indians at Fowltown, has been received. Although the necessity of this attack, and the consequent effusion of blood, is exceedingly to be regretted, yet it is hoped that the prompt measures which were taken by you on your arrival at Fort Scott, and the display of such an efficient force in that quarter, will induce the Indians to abstain from further depredations, and sue for peace.[4]

The Secretary of War reminded General Gaines of previous instructions from the department that he should avoid crossing the national boundary into Florida, but now altered those instructions somewhat to give more discretion to the commanders in the field:

Referring to the letter addressed to you from this Department on the 30th of October and 2d of December, as manifesting the views of the President, I have to request that you conform to the instructions therein given. Should the Indians, however, assemble in force on the Spanish side of the line, and persevere in committing hostilities within the limits of the United States, you will, in that event, exercise a sound discretion as to the propriety of crossing the line for the purpose of attacking them and breaking up their town.[5]

As dispatches slowly made their way between the frontier and the nation's capital, the principal chiefs of the Creek Nation convened at the town of Broken Arrow near Fort Mitchell, Alabama, to discuss the Fowltown attacks and what could be done to avoid greater conflict. They had not yet heard of the attack on Scott's party:

They unanimously expressed much regret that hostilities should have commenced between the troops under General Gaines and the Fowltown Indians, who reside within our boundary, because those Indians, although they did not unite with the friendly ones during the late war, neither did they join the Red Sticks, and had recently expressed a great desire to become decidedly friendly. They were, however, perfectly willing that their warriors should join General Gaines against the Seminoles. I stated to them that it was not the desire of the President to go to war with the Seminoles, if he could honorably avoid it....[6]

Indian Agent David B. Mitchell, the former Governor of Georgia who had replaced the well-known Benjamin Hawkins after the latter's death in 1816, told the Creeks that President Monroe would not approve of their going to war with the Seminoles on behalf of the United States and cautioned them to remain peacefully in their towns unless called upon for assistance. He did encourage them to send an emissary down the river:

I advised them to send a confidential and trusty chief down to the Indians living between Fort Gaines and the Spanish line, and desire them immediately to remove above the line of Jackson's treaty, and that the same chief should then proceed directly to the Mickasuky town, the head-quarters of the Seminoles and Red Sticks of the late war, and propose to them certain terms of peace and a junction of their force to go against the negro camp. The objects which this chief was instructed to hold out to those Indians, as attainable by adopting this course, were various, and of sufficient importance, in the view of those making the proposition, to induce a belief that they would be favorable received; in which event, I should proceed to Fort Scott to adjust their differences.[7]

Hopoi Hadjo, the head chief of Oosoochee, was selected by the council to represent the principal leaders of the Creeks to the lower towns. In order to give time for this mission to succeed as well as for him to communicate with his superiors in Washington, Mitchell requested that the assembly reconvene on January 11, 1818. At that time, per his request, a large force of the nation's warriors would assemble with William McIntosh of Coweta at their head:

But, on my return to this place, I fortunately fell in with General Gaines on his way to Fort Hawkins, from whom I learned the fatal disaster which had befallen a detachment of his troops under Lieutenant Scott, on the 30th of last month; the particulars of which he informed me he had communicated, which renders a detail from me unnecessary.[8]

General Gaines had reached Fort Hawkins on December 13th. In addition to meeting briefly with Mitchell, he found Brigadier General Thomas Glasscock at the fort with the first troops of the Georgia Militia that had responded to the call issued prior to the Fowltown incidents. The troops were in good condition, he reported to the Secretary of War, but due to "inattention on the part of the contractor's agent" did not have the supplies to march to the relief of Fort Scott.

In a dispatch penned on the 15th, Gaines warned that once the warriors of the warring towns realized they could not overwhelm the regular troops at Fort Scott, they would likely send war parties against the frontier settlements. The Seminole towns in particular, he reported, did not understand the true power of the United States:

The Seminole Indians, however strange and absurd it may appear to those who understand little of their real character and extreme ignorance, entertain a notion that they cannot be beaten by our troops. They confidently assert that we have never beaten them, or any of their people, except when we have been assisted by red people. This will appear the less extraordinary when it is recollected that they have little or no means of knowing the strength and resources of our country; they have not travelled through it; they read neither books nor newspapers; nor have they opportunities of conversing with persons able to inform them. I feel warranted, from all I know of these savages, in saying they do not believe we can beat them.[9]

As Gaines was writing his letter at Fort Hawkins, his report on the Fowltown battle of November 23rd reached both Major General Andrew Jackson in Nashville and the new Secretary of War, John C. Calhoun, in Washington. Jackson expressed hope that "this check on the savages" would prevent further hostilities, but warned the Secretary that he was not convinced:

...Should it not, and their hostility continue, the protection of our citizens will require that the wolf be struck in his den; for, rest assured, if ever the Indians find out that the territorial boundary of Spain is to be a sanctuary, their murders will be multiplied to a degree that our citizens on the southern frontier cannot bear. Spain is bound by treaties to keep the Indians within her territory at peace with us; having failed to do so, necessity will justify the measure, after giving her due

notice, to follow the marauders, and punish them in their retreat. The war hatchet having been raised, unless the Indians sue for peace, your frontier cannot be protected, without entering their country. From long experience, this result has been fully established.[10]

The War Department apparently had already reached the same conclusion. On the same day – December 16th – Secretary Calhoun authorized Jackson to cross the Florida line into Spanish territory to carry the war to the principal towns of the Seminoles:

On the receipt of this letter, should the Seminole Indians still refuse to make reparation for their outrages and depredations on the citizens of the United States, it is the wish of the President that you consider yourself at liberty to march across the Florida line, and to attack them within its limits, should it be found necessary, unless they should shelter themselves under a Spanish post. In the last event, you will immediately notify this department.[11]

The orders left no doubt that Jackson was authorized to lead U.S. troops on an invasion of a foreign country. Florida was then part of Spain and would remain so until 1821. Although many modern writers blame the general himself for his 1818 march into Florida, he clearly was following direct orders from Washington.

The instructions from the nation's capital to the military commanders became even more direct on December 26, 1817, when Gaines' report of the destruction of Lieutenant Scott's command finally reached the War Department. Secretary Calhoun immediately ordered General Jackson to the frontier and provided what information he could on the state of affairs there:

The increasing display of hostile intentions by the Seminole Indians may render it necessary to concentrate all the contiguous disposable force of your division upon that quarter. The regular force now there is about eight hundred strong, and one thousand militia of the State of Georgia are called into service. General Gaines estimates the strength of the Indians at two thousand seven hundred. Should you be of the opinion that our numbers are too small to beat the enemy, you will call on the Executives of the adjacent States for such additional militia force as you may deem requisite.[12]

The Nation Reacts

On the same day, the Secretary wrote to General Gaines assuring him that the War Department did not blame him for the deaths of Lieutenant Scott and the men, women and children under his command:

The fate of the detachment under Lieutenant Scott is much to be regretted; but, under all the circumstances, no blame can attach to yourself or the officers immediately concerned. When the order of the 12 November was given, directing you to repair to Amelia Island, it was hoped the Seminoles would have been brought to their reason without an actual use of force, and that their hostility would not assume so serious an aspect. It is now a subject of much regret, that the service in that quarter has been deprived of your well known skill and vigilance.[13]

Expressing hope that the U.S. occupation of Amelia Island would be completed by the time Gaines received his letter, Calhoun instructed him to return to Fort Scott as quickly as possible. The Secretary informed the general that Major General Jackson has been directed to assume command in person on the frontier as quickly as possible, but directed that Gaines should reassume command at Fort Scott until Jackson could reach the scene. Calhoun also proposed the possibility of a march into Florida from the east in conjunction with Jackson's anticipated movement down the Apalachicola River:

...[If] you should think the force under your command sufficient, and other circumstances will admit, to penetrate through Florida, and co-operate in the attack on the Seminoles. I am not sufficiently acquainted with the topography of the country between Amelia and their towns, to say whether it is practicable, or what would be the best route; but it is not improbable that some advantage might be taken of the St. John's river, to effect the object. Should it be practicable, it is probable efficient aid might be given to the attack on them, as the attention of the warriors must be wholly directed towards Fort Scott.[14]

Gaines did not have a sufficient force at his disposal to carry out such an invasion, but the suggestion by Calhoun provides an additional example of the thinking of the Monroe Administration regarding the movement of U.S. troops into Spanish Florida.

Even as these orders were being dispatched, Hopoi Hadjo returned from his mission to the lower towns and Miccosukee. He found the Seminoles of the later place anxious to avoid involvement in the spreading conflict. The emissary joined

85

with Tustenuggee Hopoi (Little Prince) in reporting that the Miccosukee leaders placed the blame for the destruction of Scott's party clearly on the heads of the Red Sticks, while also blaming the whites for starting the war with the attack on Fowltown:

The Mackasookies say it was not them that began the war; they were sitting down in peace, and the white people came on them in the night and fired on them. The Mackasookies are all sitting in their town and doing no mischief, and waiting to see if the white people will make peace with them. The people that shot at the boat, and killed all the white people, were the old Red Sticks from the Upper town – those that turned hostile last war. The man that was sent to the Mackasookies (Hopoie Haija) with a peace-talk met the Mackasookies at the half-way ground, coming with a peace-talk to us.[15]

The United States disregarded the attempt by the Miccosukees to make peace. The deaths of Lieutenant Scott and the men, women and children of his command so angered the leaders of the country that the invasion of Florida and destruction of its Seminole, Red Stick and African towns became the only option considered by President Monroe and Secretary Calhoun. This invasion forever changed the history of the United States and ultimately assured that Florida would become an American possession.

---

[1] *Rutland Herald*, January 7, 1818, p.3.

[2] *Massachusetts Spy*, December 31, 1817, p2.

[3] Maj. Gen. Edmund P. Gaines to the Secretary of War, December 4, 1817, *American State Papers*, Indian Affairs, Volume 2, p. 161.

[4] George Graham, Secretary of War, to Maj. Gen. Edmund P. Gaines, December 9, 1817, *American State Papers*, Indian Affairs, Volume 2, p. 161.

[5] *Ibid.*

[6] David B. Mitchell, Indian Agent, to Acting Secretary of War George Graham, December 14, 1817, *American State Papers*, Indian Affairs, Volume 2, p. 161.

[7] *Ibid.*

[8] *Ibid.*

[9] Maj. Gen. Edmund P. Gaines to the Secretary of War, December 15, 1817, *American State Papers*, Indian Affairs, Volume 2, p. 162.

[10] Maj. Gen. Andrew Jackson to the Secretary of War, December 16, 1817, *American State Papers*, Indian Affairs, Volume 2, p. 162.

[11] John C. Calhoun, Secretary of War, to Maj. Gen. Andrew Jackson, December 16, 1817, *American State Papers*, Indian Affairs, Volume 2, p. 162.

[12] John C. Calhoun, Secretary of War, to Maj. Gen. Andrew Jackson, December 26, 1817, *American State Papers*, Indian Affairs, Volume 2, p. 162.

[13] John C. Calhoun, Secretary of War, to Maj. Gen. Edmund P. Gaines, December 26, 1817, *American State Papers*, Military Affairs, Volume 1, p. 689.

[14] *Ibid.*

[15] Tustennogee Hopoi and Hopoi Haija to David B. Mitchell, Indian Agent, December 30, 1817, *American State Papers*, Military Affairs, Volume 1, pp. 692-693.

Chapter Nine

"I have taken their lives."

FROM THE MOMENT THE REPORT ON THE SCOTT BATTLE REACHED WASHINGTON, the War Department anticipated no other outcome for the war than the destruction of the Indian forces that had taken part in the attack. Militia troops were called out in Georgia, the Mississippi Territory, Tennessee, Kentucky and the Carolinas. Additional regular forces were ordered to the frontier and Generals Jackson and Gaines began their movements to Fort Scott. When the Creek leaders reconvened at Broken Arrow on January 11, 1818, thousands of their warriors were requested to prepare for a march down the Chattahoochee River to the border.

The war had not taken a lull following the destruction of Scott's party. On December 13, 1817, a war party of Red Sticks led by a secondary chief from Fowltown attacked the plantations of William Hambly and Edmund Doyle and took both men captive. William Perryman, who had gone down with his warriors from Tellmochesses, to protect the traders was killed in the fight and his surviving men forced to join the attacking force.

The attack at Spanish Bluffs was followed by simultaneous attacks on Fort Hughes and Muhlenberg's supply vessels on the Apalachicola. Captain McIntosh and his small command held off a large attacking force at Fort Hughes, but Muhlenberg was penned down in the Apalachicola for weeks as warriors rained lead shot down on his vessels from both sides of the river. The affair is remembered today as the Battle of Ocheesee. Several exchanges of fire also took place throughout December between the garrison of Fort Scott and war parties that hovered around the post.

As General Gaines had predicted, the warring towns sent parties of warriors out to strike at the frontier. Attacks on civilian targets were reported from near the St. Mary's River in Georgia all the way west to near the Alabama River in Alabama. Citizens of the region responded by "forting in" as best they could, either in strong houses or in the small forts that dotted the frontier. Homes and farms were abandoned and many went up in flames set by the war parties.

At Fort Gaines, which had been designed for two companies of regular troops, nearly 300 soldiers and civilians took refuge within the walls under miserable conditions. The situation was so overcrowded, however, that some of the families soon began returning to their homes over the advice of the fort's commander:

After all that I have said to the citizens, they are going from the fort to their houses. General Gaines directed me to send you the census of people at this post. A few days since there were two hundred and eight-five persons in the fort, sixty of which have left it. The General directed me to have a large corn-house built for the reception of the people's corn; I have done so, but they have no disposition to do so. I am constantly advising the people to secure their provisions, but they will not take advice of it till it will be too late.[1]

An attack did soon follow near Fort Gaines, leaving the mutilated bodies of several men on the ground. The people of the area flooded back into the fort, watching from its walls and blockhouses as the smoke rose from their homes and fields. Couriers sent from that post to Fort Scott turned back, often after reaching Spring Creek within just a few miles of the latter place. Likewise, no messengers from Fort Scott arrived at Fort Hawkins or Milledgeville with updates on the situation there. To officials and citizens of more populated areas alike, it was as if a wall of silence had descended over the frontier.

The troops, however, continued to assemble and by the end of January a considerable army was forming. Jackson was heading south with a body guard of Kentucky militia while regiments of Tennessee horsemen were taking the field and preparing to march for Fort Scott. In Georgia, General Glasscock's militia advanced to the Flint River and erected Fort Early near present-day Cordele. And on the Chattahoochee, McIntosh, now bearing a commission as brigadier general of volunteers, assembled a brigade of more than 2,000 Creek warriors for an advance down both sides of the river. His plan was to sweep the valley clean of Red Stick and Seminole war parties before attacking Ekanachatte, a large town in the northeast corner of what is now Jackson County, Florida, that had allied itself with the war towns.

It was McIntosh's command, in fact, that struck the first blow of revenge against the warriors responsible for the destruction of Scott's command:

...I have taken three of our enemies that were firing on the vessels on this river, and one was wounded at the same place when firing on the vessels. I have got them in strings, carrying them to Fort Gaines, and expect to catch some more before I get there.[2]

The Creek general interrogated his prisoners and quickly determined they had taken part in the attack on Lieutenant Scott's boat. His note that one of the men had been wounded in the battle is the only known reference to Indian casualties in the fight. The three captives soon paid the ultimate penalty for their involvement in the battle:

...I carried them to Fort Gaines to the commanding officer, and he told me he would have nothing to do with them, and said to me, you may deal with them by your own laws. We had proof that they were at the destroying of the boat below the fork of Flint river, and one of them was wounded at that time – they were doing mischief to our friend and I knew what was the law between us and the United States; I did not want them to stand on our land, and I have taken their lives....[3]

The first revenge for the death of Lieutenant Scott and his command, therefore, was not carried out by regular soldiers of the United States, but by Creek warriors allied with the army and under the leadership of a Coweta chief turned U.S. general. While the account does not specifically say so, General

McIntosh penned it from Fort Gaines and it appears the executions took place there as well.[4]

The Creek soldiers continued their advance down the west side of the Chattahoochee, accepting the surrenders of another 24 Red Sticks and their families before destroying Ekanachatte and capturing 53 of its men and 180 of its women and children. Another 30 of the town's warriors, led by the chief Econchattimico himself, managed to escape following a confrontation on the Chipola River, but 10 others were killed while fighting McIntosh's men.[5]

The push down the Chattahoochee by General McIntosh was the first significant U.S. counterattack since the attack on Lieutenant Scott the previous November. It also signaled the beginning of the retaliatory campaign planned by American commanders to exact revenge for the death toll in the attack and punish those responsible for it.

---

[1] Capt. Robert Irvin to Lt. Col. Matthew Arbuckle, December 23, 1817, *American State Papers*, Military Affairs, Volume 1, p. 692.

[2] Brig. Gen. William McIntosh to Maj. Daniel Hughes, March 2, 1818, *Camden Gazette*, April 11, 1818, p. 3.

[3] Brig. Gen. William McIntosh to Maj. Daniel Hughes, March 5, 1818, *Camden Gazette*, April 11, 1818, p. 3.

[4] *Ibid.*

[5] Brig. Gen. William McIntosh to Maj. Daniel Hughes, March 10 & 16, 1818, *Camden Gazette*, April 11, 1818, p. 3.

Chapter Ten

Jackson Invades Florida

AS WILLIAM McINTOSH'S COMMAND WAS MAKING ITS SUCCESSFUL ADVANCE DOWN THE CHATTAHOOCHEE RIVER, the main body of Jackson's army pushed southwest down the Flint. Accompanying the general was the father of Elizabeth Stewart. Her husband, meanwhile, awaited the army's arrival at Fort Scott.

Jackson's advance was hampered more by chronic supply shortages and cold weather than it was by Red Sticks and Seminoles. In fact, the various accounts of the march from Fort Early to Fort Scott mention no fighting at all. They do report that the winter had turned severely cold. Private John Banks of the Georgia Militia, for example, noted that the ponds and streams were choked with ice and described a March snow that fell deep into Southwest Georgia.[1]

The effects of the Tambora eruption continued to disturb the normal weather patterns of the world and the "Year Without a Summer" stretched into two years of abnormal cold in the Deep South. The cold likely had much to do with the success of the advances by McIntosh and Jackson. After maintaining a running battle with Muhlenberg's supply vessels on the Apalachicola for weeks, the warriors of the Red Stick/Seminole alliance had suddenly melted away and returned to their towns. Spring found them still huddled there, hovering around their fires and facing shortages of food and ammunition. William Hambly and Edmund Doyle were carried to Boleck's town on the Suwannee and often threatened with death, but the Black Seminole chief Nero intervened on their behalf and they were turned over to the Spanish commandant at San Marcos de Apalache (Fort St. Marks) for safekeeping. During their time there, they had good opportunity to observe the meetings of the principal leaders of the war effort:

Whilst at the port the ingress and egress of the Indians, hostile to the United States, was unrestrained, and several councils were held, at one of which Kenhajah, King of the Mickasukians, Francis or Hillis Hago, Hamathlemeco, the chief of Autesses, and the chief of the Kolemies, all of the old Red Stick party; and Jack Mealy, chief of the Ochewas, were present. When it was reported that the chiefs, and that warriors were entering Fort St. Marks for the purpose of holding a council, Hambly represented to the commandant the impropriety of permitting such proceedings with the walls of a Spanish fortress, the officer of which was bound to preserve and enforce the treaties existing between the King of Spain and the United States; he replied to Hambly with some degree of warmth, observing that it was not in his power to prevent it. On the Indians coming into the fort, at their request, we were confined. The council was held in the commandant's quarters, he, the commandant, was present, but strictly forbade the intrusion of any of the officers of the garrison.[2]

The Spanish commandant later defended himself against American charges that he had conspired with the chiefs of the hostile alliance by using the same excuse, that his command was so small he could not have resisted the warriors of the Red Stick/Seminole alliance had he so desired. U.S. officials, of course, disagreed and the role of the Spanish in assisting the war effort would play a key part in the post-war negotiations between the two nations.

Andrew Jackson reached Fort Scott in person on March 9, 1818. The situation at the fort was perilous. Food supplies were running severely short and the general brought with him 900 Georgia militiamen and several hundred more friendly Creek warriors, but only 1,100 head of live hogs to feed the whole. Contractors had failed to deliver expected provisions to Fort Scott and the general found there only a few head of poor cattle and a small quantity of corn to feed both the regulars already at the post as well as the soldiers he had brought with him. The general's Creek War experience had taught him that to halt even temporarily in hopes that supplies might arrive could mean the destruction of his army. Accordingly, he assumed command at Fort Scott the next morning, ordered the livestock slaughtered and at 12 noon began crossing the Flint River with a total force of more than 2,000 men.[3]

It was a desperate move. The distribution of all but a meager supply of provisions being left for the 60 regulars assigned to stay behind and garrison the fort gave each man in the army only one quart of corn and a three-day ration of meat, primarily raw pork. The only hope was that supply vessels expected from New Orleans were on their way up the Apalachicola River and that Jackson might meet them somewhere between Fort Scott and the bay:

...Having to cross the Flint river, which was very high, combined with some neglect in returning the boats during a very dark night, I was unable to move from the opposite bank until nine o'clock, on the morning of the 11th, when I took up my march down the each bank of the river...touching the river as often as practicable, looking for the provision boats which were ascending, and which I was fortunate enough to meet on the 13th, when I ordered an extra ration to the troops, they not having received a full one of meal or flour since their arrival at Fort Early.[4]

The army had completely exhausted its supply of meat on the 12th and the men were down to only a few handfuls of corn when the supply boats providentially appeared around the bend south of Alum Bluff in what is now Liberty County, Florida. Private Banks described in his diary how he immediately sat down and consumed the handful of corn he had set aside as his last resort.[5]

On the same day, as his men camped at Alum Bluff enjoying their first full meal in weeks, Jackson's sentries captured three warriors who either stumbled into or were watching the advance of the army. They also uncovered a small amount of corn hidden in the vicinity. Two evenings later, on the 15th, one of the prisoners tried to escape, igniting near chaos in the camps:

...On Sunday night a little after sunset, as the army was in the attitude of stacking arms, and all was in confusion, the hostile Indian endeavored to make his escape. He was fired on by about fifty of the guards and killed. Those who were not contiguous to the scene thought the army was attacked by the enemy. It produced great confusion in the ranks.[6]

The army reached Prospect Bluff, site of the ruins of the "Negro Fort," on March 16, 1818. The water battery of the fortification was still largely intact, as were the outer ditches and breastworks. The magazine, scene of the explosion that destroyed the fort on July 27, 1816, was just a crater surrounded by mounds of earth. The soldiers explored the desolate scene, finding weapons and other artifacts from the explosion still lying in the dirt:

...We found some of their arms and cannon ball lying in the mud. The guns, although they had been exposed to the weather for four years, when put in the fire to burn the rust off, would fire. Here we erected a new fort upon the ruins of the old one and called it Fort Gadsden.[7]

Jackson was so pleased with the initiative and efforts of his engineer, Lieutenant James Gadsden, that he named the fort in his honor. The young officer would later achieve distinction in America's diplomatic history for negotiating the famed Gadsden Purchase from Mexico. That agreement resulted in the U.S. acquisition of a large tract of land that today comprises much of southern New Mexico and Arizona. Gadsden County, Florida, also bears his name.

On March 23rd, after several days of growing apprehension, the army received news that three supply ships and their armed escort had arrived in Apalachicola Bay. The vessels also brought news that caused Jackson to believe the Spanish fort of San Marcos de Apalache likely had been surrendered to the chiefs of the alliance warring against the United States:

...The Governor of Pensacola informed Captain Call, of the 1st infantry, (now here,) that the Indians had demanded arms, ammunition, and provisions, or the possession of the garrison of St. Marks of the commandant, and that he presumed possession would be given from inability to defend it. The Spanish Government is bound by treaty to keep her Indians at peace with us. They have acknowledged their incompetency to do this....[8]

This intelligence in hand, Jackson began preparing his army to march. His plan was clear, to strike at the principal Seminole villages of Tallahassee Talofa and Miccosukee then turn south to the Gulf and seize San Marcos de Apalache:

...[S]hould I be able, I shall take possession of the garrison as a depot for my supplies, should it be found in the hands of the Spaniards, they having supplied the Indians; but if in the hands of the enemy I will possess it, for the benefit of the

United States, as a necessary position for me to hold, to give peace and security to this frontier, and put a final end to Indian warfare in the South.[9]

The aftermath of the attack on Lieutenant Scott's party now began to take on even more serious international implications. Additional supplies reached Fort Gadsden on March 24th and the army broke camp the next morning and began its march for Tallahassee and Miccosukee. The route of the march took Jackson back to the north and then eastward around the morass of the Tate's Hell Swamp. He re-entered present-day Gadsden County and reached the Ochlocknee River on the 29th at a point opposite the Leon County bluff that today bears his name.

The river was running high, but the aggressive general ordered his men to begin building canoes and on March 30, 1818, the 2,000 man army crossed the Ochlockonee and on the next day reached the town of Tallahassee Talofa:

...On Tuesday, the 31st, arrived at a town called Tallahassee. The Indians had abandoned it before we got there. We passed an old Indian lying near a pond, dead. She had not been dead long from her appearance; she had been left there to die by the Indians who fled before us; she was lying on the ground by some ashes and a dirt pot. We burnt the town (this is the present capital of Florida). We found some cattle that day which were distributed amongst the soldiers.[10]

The arrival of the main army at Tallahassee had been preceded by the rapid push forward of a force under Major Twiggs. Intelligence had been received that the band of Red Sticks holding Elizabeth Stewart was at the village and General Jackson attempted to attack them before they could escape with their captive:

...We knew long before we re-captured her what band she was with, and had tried to come up with them before...The most tiresome march I ever made was one night in company with the present Gen. Twiggs. He with some soldiers, and I with a party of Indians, trying to rescue her at old Tallahassee, but the Indians had left before we reached the place.[11]

After camping that night near the abandoned Seminole town, Jackson advanced the next morning for Miccosukee, the principal seat of the Seminole Nation. The Creek warriors under General McIntosh joined the army that morning, having arrived at Fort Scott after the departure of the main army and then marching cross-country to find and reinforce Jackson. A portion of the long-awaited Tennessee Mounted Volunteers also appeared that same morning, joining the heavily reinforced army as it marched forward:

...On the same day, a mile and a half in advance of the Mekasukean villages, a small party of hostile Indians were discovered judiciously located on a point of land projecting into an extensive marshy pond; the position designated, as since understood, for concentrating the negro and Indian forces to give us battle. They sustained for a short period, a spirited attack from my advanced spy companies; but fled and dispersed in every direction, upon coming in contact with my flank columns, and discovering a movement to encircle them.[12]

The army had apparently arrived faster and in greater force than Cappachimico had expected. His fleeing warriors were pursued through the vast Miccosukee villages, which extended for miles along the western shore of Lake Miccosukee. The largest Seminole town in Florida and the seat of power for the western division of that nation since the days of the American Revolution fell to Andrew Jackson in a chaotic running battle that lasted minutes, not hours.

As Jackson and his officers rode into the villages and approached the council house and home of Cappachimico, they discovered a grisly sight:

...In the council houses of Kenhagees town, the King of the Mekasukians, more than fifty fresh scalps were found; and in the centre of the public square, the old Red Stick's standard, a red pole, was erected, crowned with scalps, recognized by the hair, as torn from the heads of the unfortunate companions of Scott.[13]

Whether or not the Miccosukee warriors had participated in the attack on Scott's command, participants in the battle had taken their trophies to the town for display there. Private John Banks of the Georgia Militia also described the scalp pole, where he reported seeing many scalps nailed. "One could discover from the length of the hair," he wrote, "several female ones."[14]

The sobering sight reminded the soldiers of the purpose of their march and instilled in their commander an even more firm belief in the righteousness of his actions. He sent General Gaines with a large command the next day to wade across Lake Miccosukee and destroy the new town built by Neamathla and the Fowltown band. This was accomplished without significant resistance. Other soldiers spread out through the Miccosukee villages, collecting cattle, corn and supplies. Then, on April 5, 1818, the torch was applied to the town and hundreds of homes destroyed. The army marched away for St. Marks with the horizon behind it punctuated by the smoke rising from the burning homes of Cappachimico and his people.

---

[1] John Banks, *Diary of John Banks*, 1936, pp. 9-14.

[2] William Hambly and Edmund Doyle to "Sir," May 2, 1818, *American State Papers*, Military Affairs, Volume 1, p. 715.

[3] Maj. Gen. Andrew Jackson to John C. Calhoun, Secretary of War, March 25, 1818, *American State Papers*, Military Affairs, Volume 1, pp. 698-699.

[4] *Ibid.*

[5] John Banks, *Diary of John Banks*, 1936, pp. 9-14.

[6] *Ibid.*

[7] *Ibid.*

[8] Maj. Gen. Andrew Jackson to John C. Calhoun, Secretary of War, March 15, 1818, *American State Papers*, Military Affairs, Volume 1, pp. 698-699.

[9] *Ibid.*

[10] John Banks, *Diary of John Banks*, 1936, pp. 9-14.

[11] Thomas Woodward to John Banks, June 16, 1858.

[12] Maj. Gen. Andrew Jackson to John C. Calhoun, Secretary of War, April 8, 1818, *American State Papers*, Military Affairs, Volume 1, pp. 699-700.

[13] *Ibid.*

[14] John Banks, *Diary of John Banks*, 1936, pp. 9-14.

Chapter Eleven

Execution of Francis and Homathlemico

BY THE TIME ANDREW JACKSON INVADED FLORIDA, the blame for the attack on Lieutenant Richard W. Scott and his command had been placed primarily on the warriors of the Red Stick chief Homathlemico. A fierce opponent of the United States, he had led his people south into Spanish territory after Jackson's bloody victory at Horseshoe Bend. A close associate of Peter McQueen and the Prophet Josiah Francis, Homathlemico and his warriors had supported Lieutenant Colonel Edward Nicolls and the British during the War of 1812. When Nicolls evacuated the Apalachicola River at the end of that conflict, the chief remained in the "Big Bend" region of Florida, determined that he would never again be driven from his home.

When news of the Fowltown attacks reached the Florida villages, the Red Stick chief and his men were among the first to respond to Neamathla's call for help. Wisely recognizing that victory in the war hinged on their ability to block the delivery of supplies to the troops at Fort Scott, Homathlemico and Josiah Francis moved with their warriors and any others they could collect to the banks of the Apalachicola. While Francis is believed to have exercised overall command of the large army that gathered in the vicinity of Ocheesee Bluff, Homathlemico led the force that assembled near the forks of the Chattahoochee and Flint Rivers.

Homathlemico and Francis were at San Marcos de Apalache (Fort St. Marks) when news reached them that Jackson's Army was marching through the country near Tallahassee Talofa and Miccosukee. Their situation was dire as they were in desperate need of ammunition to oppose the soldiers. The adventurers Alexander Arbuthnot and Robert Ambrister had promised them help should the need arise and when they saw a ship sail into the St. Marks River flying the flag of Great Britain, they believed their long sought relief had arrived.

The two chiefs paddled out from the old Spanish fort to greet the captain of the vessel, but when they climbed aboard it only took seconds for them to realize they had been decoyed into a trap. The ship was not a British supply vessel, but the USS *Thomas Shields,* an American warship captained by Lieutenant Isaac McKeever of the U.S. Navy:

...Captain McKeever, of the navy, having sailed for St. Marks with some vessels containing supplies for the army, was fortunate enough to entice on board his vessel, in the river, Francis, or Hillis-Hajo, and Homathlemicco, hostile chiefs of the Creek nation, and whose settled hostility has been severely felt by our citizens. The commanding general had them brought on shore, and ordered them to be hung, as an example to deter others from exciting these deluded wretches to future scenes of butchery.[1]

The hanging of Francis and Homathlemico was quick and without trial or ceremony. Jackson had no love lost for either man, but Francis in particular had been a thorn in his side for years. He was the prophet who had ignited the spread of the Shawnee Prophet's new religion among the Creeks. From his town of Holy Ground on the Alabama River, Josiah Francis had taught the new religion first to a small band of converts but in a remarkably short time lit a fire that raged through the Upper Creek towns. His following grew in numbers from a few to dozens, then hundreds and finally thousands. The teachings of the prophet were not as warlike as many writers since his day have supposed. Instead, he advocated a complete separation of the Creeks from the influence of the whites. Warriors and their families, he taught, should return to their traditional ways and give up all white goods with the exception of their weaponry. They should concentrate on their families and work hard to obtain the approval of the Master of Life. Indians, he taught, should give up alcohol and other bad things and should never steal from anyone. They should live in peace with the whites if possible, but under no circumstances should they ever consent to give up another inch of land to any other nation.

While his abilities as a military leader often were ridiculed by later writers, Francis actually fought Andrew Jackson to a standstill at the Battles of Emuckfau and Enitichopco. He and McQueen similarly battered the Georgia army under Major General John Floyd at the Battle of Calabee Creek. British sources, in fact, placed Francis on the scene of the successful Red Stick attack on Fort Mims, the bloody battle for revenge that brought the United States formally into the Creek War of 1813-1814. While American sources of the 19th century downplay his influence and abilities, he clearly was a charismatic and capable leader who at the height of his power commanded thousands of warriors.

Jackson had desperately tried to capture Francis at the end of the Creek War, but the prophet eluded pursuit and fled with his family and core of followers across the line into Spanish Florida. He allied himself with the British upon their arrival on the Gulf Coast and was reported as present in British uniform at the forts on the Apalachicola in early 1815. In attendance at the conference of Red Stick and Seminole leaders that convened at the British outpost near the confluence of the Chattahoochee and Flint Rivers in March 1815, Francis was designated as a representative of the alliance and sent to England with Colonel Nicolls to solicit continued British support. He returned from England in the spring of 1817, having met with the Prince Regent and other Royal officials. In fact, he even left behind his young son who remained at the home of Colonel Nicolls, who promised to see to the boy's education. After his arrival back in Florida aboard the schooner of Alexander Arbuthnot, the prophet seems to have provided a voice of moderation among the towns of the North Florida and South Georgia region. When war finally did erupt, he led warriors into action at the Battle of Ocheesee, but the real evidence indicates that he had little to do with the outbreak of the conflict.

Jackson, of course, had no concern about or regard for such changes in Francis and his teachings. McKeever's deception had lured the old enemy of the Americans into their hands and the general wasted no time in ordering his immediate execution.[2]

Not as well known in his day as the Prophet Francis, Homathlemico was equally of interest to General Jackson and the other U.S. commanders. Little is known of his movements and activities between the end of the War of 1812 and the outbreak of the First Seminole War. He appears to have been among the chiefs who established themselves at or near Miccosukee, probably due to his close association with Francis and McQueen who settled nearby on the Wakulla River. The Spanish fort at St. Marks also was an important point for supply and trade.

Captured alongside Francis by Lieutenant McKeever, Homathlemico met his fate at the same time as the prophet:

...While passing 'round, the commander managed to decoy some Indians, who made to the vessel under the impression that it was a British sloop. Two Seminolean chiefs were taken (Prophet Francis and King Michelimico). On Wednesday, the 8th, both were brought from the vessel to the gallows and hung without ceremony or trial.[3]

The site of the execution appears to have been in the level area just outside the gates of the Spanish fort. The San Marcos de Apalache Historic State Park museum stands in this area today, as does a monument to Milly Francis, a daughter of the Prophet Francis. Not long before the capture of her father, Milly had saved the life of a Georgia militiaman named Duncan McCrimmon (usually spelled McKrimmon). Captured by warriors loyal to Josiah Francis, he had been brought to the prophet's town on the Wakulla River and was about to be executed when Milly intervened on his behalf and convinced the warriors to spare him. McCrimmon was turned over to the commandant at San Marcos and, so far as is known, did not attempt to intervene to save the life of the father of his rescuer. Milly is remembered in story and legend today as the "Creek Pocahontas."

By the time Josiah Francis and Homathlemico were brought ashore, of course, Jackson's troops had occupied the fort of San Marcos de Apalache. Its Spanish garrison made a feeble attempt to defend their position, but the defenses were taken without the firing of a shot by either side. Inside the fort, the soldiers found Alexander Arbuthnot. He had come to St. Marks in an attempt to save his trade goods and other possessions, but was trapped there when McKeever's vessel suddenly appeared in the river and Jackson's army marched down from Miccosukee:

In Fort St. Marks, as an inmate in the family of the Spanish commandant, an Englishman by the name of Arbuthnot was found, unable satisfactorily to explain the objects of his visiting this country, and there being a combination of circumstances to justify a suspicion that his views were not honest, he was ordered in close confinement.[4]

To Jackson, who harbored a lifelong dislike of the British, the discovery of Arbuthnot in the fort that he believed was the primary point of supply for the Indian alliance only strengthened his conviction that foreign agents and adventurers had instigated the war. The elderly Scotsman would be tried before a military tribunal for his role in the war.

Having placed Arbuthnot in confinement, Jackson again ordered his men to prepare to march. Their next target was Boleck's town on the Suwannee River. The general knew that the last large group of organized Seminole warriors would be found there and, of long interest to the Americans, so to would be found the village of the African or Black Seminoles. They lived in a town near but under the protection of Boleck. Rumored to have played an important role in the war thus far, their destruction was considered a key military objective of the campaign.

Behind them as they marched away, the soldiers left the bodies of Josiah Francis and Homathlemico. Undoubtedly tossed into a hastily and unceremoniously dug hole, the exact site of the grave of the Indian leaders is not currently known. Somewhere on or near the grounds of San Marcos de Apalache Historic State Park, however, two of the most significant figures of the 19th century rest all but forgotten in an unmarked grave. Francis had paid the price for his role in igniting the Creek War of 1813-1814. For Homathlemico, the Americans considered his death at the end of the rope to be a just penalty for the deaths of Lieutenant Richard W. Scott, the soldiers of his command, six women and four children:

...[Francis] has since signalized himself by his murders, and depredations; and has been a principal cause of exciting the Seminoles to hostility and provoking the present contest. The other [Homathlemico], who suffered with him the just punishment of their crimes, was the fellow who commanded the savages at the murder of lieut. Scott and his party on Flint river. We think general Jackson deserves commendation rather than censure, for having refused to such wretches, the rights of prisoners of war.[5]

The Red Stick chief was the fourth and final person to be executed for his role in the attack on Scott's command. The campaign instigated by the encounter, however, was far from over.

[1] Robert Butler, Adjutant General, to Brig. Gen. Daniel Parker, Adjutant and Inspector General, May 3, 1818, *American State Papers*, Military Affairs, Volume 1, pp. 703-704.

[2] Maj. Gen. Andrew Jackson to John C. Calhoun, Secretary of War, April 9, 1818, *American State Papers*, Military Affairs, Volume 1, p. 700.

[3] John Banks, *Diary of John Banks*, 1936, pp. 9-14.

[4] Maj. Gen. Andrew Jackson to John C. Calhoun, Secretary of War, May 5, 1818, *American State Papers*, Military Affairs, Volume 1, pp. 701-702.

[5] *Kentucky Reporter* quoted in the *Weekly Aurora*, June 22, 1818, Volume IX, Issue XVIII, p. 139.

Chapter Twelve

The Rescue of Elizabeth Stewart

THE NATURAL BRIDGE OF THE ECONFINA RIVER is a beautiful and pristine place. Some might call it remote, but it is only a 20-minute drive from Perry, Florida. Tallahassee, meanwhile, is only 45 minutes away, off to the northwest. U.S. Highway 98 passes just a short distance north of the unique natural feature.

Only a real "natural bridge" during dry weather, the Natural Bridge of the Econfina is a place where part of the river's current flows underground for a short distance before rising back to the surface. At normal water levels, the river flows over the natural bridge as well as beneath it and the old crossing place is recognizable only as a shallow place and small waterfall. At high water, even the waterfall disappears beneath the smooth and powerful flow of the Econfina.

Now preserved at the Suwannee River Water Management District's Natural Well Branch tract, the natural bridge was used as a crossing place for hundreds of years.. The trail used by Jackson's army to cross the Econfina in 1818 is still quite visible today. Somewhere nearby is the battlefield where Elizabeth Stewart was freed from her captors.

Having escaped Jackson's advance on Miccosukee and his subsequent capture of Fort St. Marks, Peter McQueen led his large band of Red Sticks eastward in a slow move for the Suwannee River. He apparently planned to join his warriors with the men of Boleck's band and the several hundred Black Seminole warriors who lived nearby. Together they would have sufficient strength to wage a rearguard action as the women, children, supplies and livestock were evacuated to safety. It was a desperate but slow movement, with the chief and his warriors encumbered by their families, possession, cattle and everything else they owned.

Among the women guarded by McQueen's warriors was the only female survivor of the Scott Battle, Elizabeth Stewart. How she came to be there is not clear. Some of the chief's men could have participated in the attack on Scott's command, or McQueen could have welcomed followers of Homathlemico into his band after their chief was captured at St. Marks. Also, with the women of the slow moving band was a sister of Peter McQueen. She had once been married to an English trader named Powell and her son, a young boy rapidly becoming a man ahead of his time, was then known by the name Billy Powell. He was around nine years old in the spring of 1818, but in later years grew to become the famed Seminole warrior, Osceola.

McQueen's town had been somewhere near that of the Prophet Francis on the Wakulla River and he had been fortunate that he was not with the latter individual when Lieutenant McKeever sailed the *Thomas Shields* into the St. Marks River. Instead, learning of the attack on Miccosukee and almost simultaneous capture of the Creek prophet, McQueen began a desperate effort to get his people to safety.

The rivers and creeks were running high and the movement was painfully slow. The halt of Jackson's army at Fort St. Marks likely gave the chief hope that his effort would succeed. The forces of the American general, however, were converging on him faster than the old Red Stick could move his followers out of harm's way.

Having occupied Fort St. Marks with a strong garrison, Jackson ordered his men to prepare to march and as the soldiers headed out across the marshes that surrounded the fort on the morning of April 9, 1818, they numbered over 2,000 muskets. To the men watching from the ramparts, the army must have presented an impressive sight, its long columns of blue and brown crossing the open marsh and disappearing into the trees.

Jackson followed the old trail that led from St. Marks west to the Suwannee. It looped in a curving path to the northeast around the worst of the coastal lowlands before crossing the Natural Bridge of the Econfina and passing on via a more direct route to the crossing on the Suwannee River near Boleck's town. There were no settlements along the way and the march was through one continuous wilderness.[1]

Apparently alerted as to the general's plan, other commands still moving to reinforce him angled their marches to form junctions with the army along its line of march. On April 10th two full regiments of mounted Tennessee volunteers made contact with Jackson and joined his column. A few regulars also arrived and then, later in the day, Brigadier General McIntosh finally reached the army with his brigade of 1,100 Creek warriors. They had been left behind at Miccosukee to scour the countryside for enemy warriors and enemy cattle. With the arrival of these two forces, the size of the army doubled in a single day. General Jackson now had a force more than capable of sweeping aside any resistance that could be offered by the remaining warriors of the Red Stick and Seminole alliance.[2]

The soldiers and officers of the army had known for some time that the Indian party holding Mrs. Stewart was just ahead of them. The general had ordered Major Twiggs to move against Tallahassee Talofa ahead of the main army after hearing that the unfortunate woman was being held there, but the Seminoles were able to evacuate the town before the major's force arrived. Somewhere in the vicinity of St. Marks the troops again picked up information on Stewart's presence among the Indians known to be withdrawing towards the Suwannee. Jackson knew he was closing in on McQueen, but no one in the army knew the exact whereabouts of the Red Stick chief until the morning of April 12, 1818:

...On the morning of the 12th, the officer of the day reported that the sentinels had heard the lowing of cattle and barking of dogs during the night; from which the general was indused to send a runner to General McIntosh, who was encamped a short distance in rear of the army, with instructions to have the country below examined. In the meantime, the army moved slowly in advance.[3]

As Jackson's main force pushed forward and across the Natural Bridge of the Econfina, McIntosh moved his command into the woods south of the trail and began a search for the source of the sounds heard during the night. Major Noble Kenard (or Kinnard) was ordered to advance deeper into the woods and find the location of what was believed to be the Red Stick camp. In short order Kinnard sent back a runner with word that he had discovered a large enemy force and needed immediate reinforcement:

...McIntosh moved against them with his whole force. A small detachment of different companies of Tennessee volunteers, under Colonels Dyer and Williamson, (they having joined the army on the evening of the 10th,) were left at our encampment to search for horses, and, on hearing the report of Major Kanard, formed themselves into a company under Captain Bell, who was with them, and moved to attack the enemy, whom they found near a large swamp endeavoring to move off.[4]

The pursuit of Peter McQueen finally came to an end. When McIntosh attacked, the chief was trying to get the women and children and large cattle herd to safety. As the Creek warriors and Tennessee volunteers appeared in the woods behind him, McQueen ordered his warriors into action. The size of the force at his command was no match for McIntosh's brigade, but most of his warriors had fought in some of the bitterest actions of the Creek War of 1813-1814. The chief and his men were seasoned and determined fighters:

...McIntosh with a part of his warriors, attacked a party of hostile Indians. The engagement continued about two hours with much spirit, when the hostiles retreated, leaving their women, children and property of all kinds to the mercy of the conquerors – Sustaining a loss of 37 killed on the field, and two wounded, and a number of prisoners. The number of Indians engaged was differently represented by different prisoners. McIntosh had three men killed, and several wounded.[5]

In the midst of the battle, as war cries, gunfire and smoke filled the swamp along the margin of the Econfina River, a woman could be heard calling for help. Elizabeth Stewart had been found:

...Shortly after the firing commenced, we could hear a female voice in the English language calling for help, but she was concealed from our view. The hostile Indians, though greatly inferior in number to our whole force, had the advantage of the ground, it being a dense thicket, and kept the party that first attacked at bay until Gen. McIntosh arrived with the main force. McIntosh, though raised among savages, was a General; yes, he was one of God's make of Generals. I could hear his voice above the din of firearms — "Save the white woman! Save the Indian women and children."[6]

Mrs. Stewart had taken advantage of the confusion created by the sudden attack to escape from those holding her and find shelter among the thick palmetto that grew on the margin of the swamp. As the battle intensified and McIntosh's main force came up, she began to call for help.

There are some discrepancies about what happened next. Thomas Woodward, who described the Battle of Econfina some 40 years later, claimed that he was directed by General McIntosh in person to go with a few other men and save the woman:

...All this time Mrs. Stuart was between the fires of the combatants. McIntosh said to me, "Chulatarla Emathla, you, Brown and Mitchell, go to that woman." (Chulatarle Emathla was the name I was known by among the Indians.) Mitchell was a good soldier and a bad cripple from rheumatism. He dismounted from his horse and said, "Boys, let me lead the way." We made the charge with some Uchees and Creeks, but Mitchell, poor fellow, was soon left behind, in consequence of his inability to travel on foot.[7]

Woodward's account was written in a letter to John Banks, the former Georgia militia private who also left a firsthand account of Jackson's campaign. Both men were present at the time, although Banks was with the main column while Woodward was on the battlefield itself. Woodward probably magnified his own importance along with that of his friends in his version – military records indicate the detachment that saved Mrs. Stewart was led by Major Noble Kinnard – but his account is the only known firsthand description of her rescue:

...I can see her now, squatted in the saw palmetto, among a few dwarf cabbage trees, surrounded by a group of Indian women. There I saw Brown kill an Indian, and I got my riflestock shot off just back of the lock. Old Jack Carter came up with my horse shortly after we cut off the woman from the warriors. I got his musket and used it until the fight ended. You saw her (Mrs. Stuart) when she reached the camp, and recollect her appearance better than I can describe it.[8]

It is to be regretted that neither Woodward nor Banks thought it of value to write down their memories of Mrs. Stewart's appearance or statements at the time of her rescue.

As the battle intensified, the heavy firing could be heard by the soldiers of Jackson's main column, which had moved forward about six miles from the scene of the action. McQueen's warriors were outnumbered, but waged the fiercest battle of the 1818 campaign. They also paid heavily in blood:

...There seemed to be a considerable number collected there. When we first began to fight them, they were in a bad swamp, and fought us there for about an hour, when they ran, and we followed them three miles. They fought us in all about three hours. We killed 37 of them, and took 98 women and children and six men prisoners, and about 700 head of cattle and a number of horses, with a good many hogs and some corn. We lost three killed and had five wounded.[9]

The nature of the fight and difficulty of the ground made it impossible for McIntosh and his officers to accurately estimate the strength of the Red Stick force. The prisoners they captured indicated to them, however, that numerous towns were represented on the field:

...Our prisoners tell us that there was 120 warriors from six different towns. From what we saw I believe there was more than they say, as some of our prisoners say there was 200 of them. Tom Woodward and Mr. Brown, and your Son, our Agent, and all the white men that live in our country, were with us through the whole fight, and fought well. All my officers fought so well I do not know which is the bravest. They all fought like men and run their enemies.[10]

The report by General McIntosh was addressed to Indian Agent David B. Mitchell and it was his son, William "Billy" Mitchell, to whom the Creek leader referred in the account. His written "talk", as he described it, also confirmed the presence of Major Woodward in the fight. As he continued his report, the general also referred to the rescue of Elizabeth Stewart and the presence of her father and spouse with the army at the time of the battle:

...There was among the Hostiles a woman that was in the boat where our friends the white people were killed on the river below Fort Scott. We gave her to her friends – her husband and father are with general Jackson – Major Kinnard took her himself.[11]

The disparity in the numbers killed on the two sides at the Battle of Econfina is difficult to comprehend, as is the fact that killed Red Stick warriors outnumbered wounded ones by a wide margin. If the figures given by the various writers associated with the army are accurate, then McQueen lost 37 men in the battle while McIntosh lost only three. Even more curious, the reported number of Red Stick dead outnumbered the total of McQueen's wounded by more than 18 to 1. While it is certainly possible that more of the chief's men were wounded but escaped, there can be but little doubt that a battle every bit as fearful as the one carried out on Scott's command took place in the swamps along the Natural Well Branch of the Econfina River.

The actual number of Red Stick warriors that escaped from the battlefield is not known. Peter McQueen did make it off the field, as did at least 100 more of the men fighting under his command. His young nephew Billy Powell, later known as Osceola, was among the women and children who escaped. When all was said and done, however, the Red Stick commander lost ten times as many men killed as did McIntosh. The Red Sticks may have been, and probably were, low on ammunition. They also might have waged a particularly ferocious delaying action against an overwhelming force to allow time for more of the women and children to escape. Whatever the situation, one thing is clear: McIntosh and his Creek warriors did not leave many Red Sticks wounded and breathing on the field at Econfina.

The Battle of Econfina finally ended. After a running fight of three miles and some three hours of constant firing, McQueen extricated what remained of his band from the engagement and slipped away into the swamps. Behind, he and his men left nearly 100 of their women and children, along with almost all their livestock and food supplies. The coming summer and winter would be a time of great suffering for the survivors. Lieutenant Scott and his command had been avenged.

---

[1] Capt. Hugh Young, "A Topographical Memoir of East and West Florida with Itineraries," 1818, National Archives.

[2] Report dated Milledgeville, Georgia, May 5, 1818, citing an officer direct from the army, published in the *Weekly Aurora*, May 25, 1818, p. 111.

[3] Maj. Robert Butler, Adjutant General, to Brig. Gen. Daniel Parker, Adjutant & Inspector General, May 3, 1818, *American State Papers*, Military Affairs, Volume 1, p. 703.

[4] *Ibid.*

[5] Report dated Milledgeville, Georgia, May 5, 1818, citing an officer direct from the army, published in the *Weekly Aurora*, May 25, 1818, p. 111.

[6] Gen. Thomas Woodward to Col. John Banks, June 16, 1858.

[7] *Ibid.*

[8] *Ibid.*

[9] Brig. Gen. William McIntosh to David B. Mitchell, Agent for Indian Affairs, April 13, 1818, *Berks and Schuylkill Journal*, May 16, 1818, p. 2.

[10] *Ibid.*

[11] *Ibid.*

Illustrations

Chapters Seven - Twelve

Restored Blockhouse at Fort Gaines, Georgia

Site of Ekanachatte (Red Ground) in Jackson County, Florida

Andrew Jackson late in life (Brady Photograph from the National Archives)

Alum Bluff in Liberty County, Florida

118

Historical Marker at Prospect Bluff Historic Sites

Earthworks of Fort Gadsden

1823 Vignoles Map showing trails followed by Jackson's Army

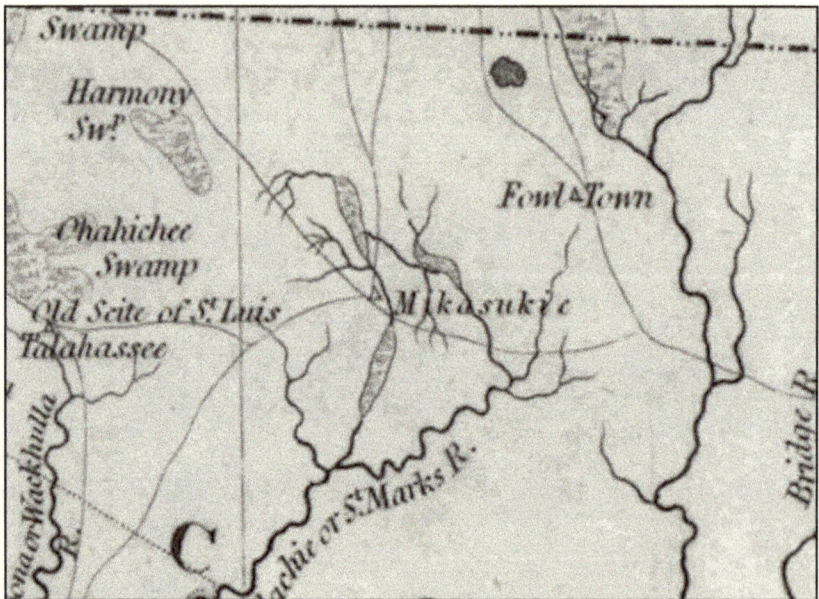

Fowltown and Miccosukee in 1818, after Neamthla's withdrawal to Florida

Photographs

Lake Miccosukee (U.S. troops waded the center of the lake)

Miccosukee Towns were on West Side of Lake (Left of Photo)

Tuckose Emathla or John Hicks, Chief of Miccosukee after death of Cappachimico.
(Probable participant at the Battle of Miccosukee in 1818)

Milly Francis begs for McCrimmon's life in this 19th century image

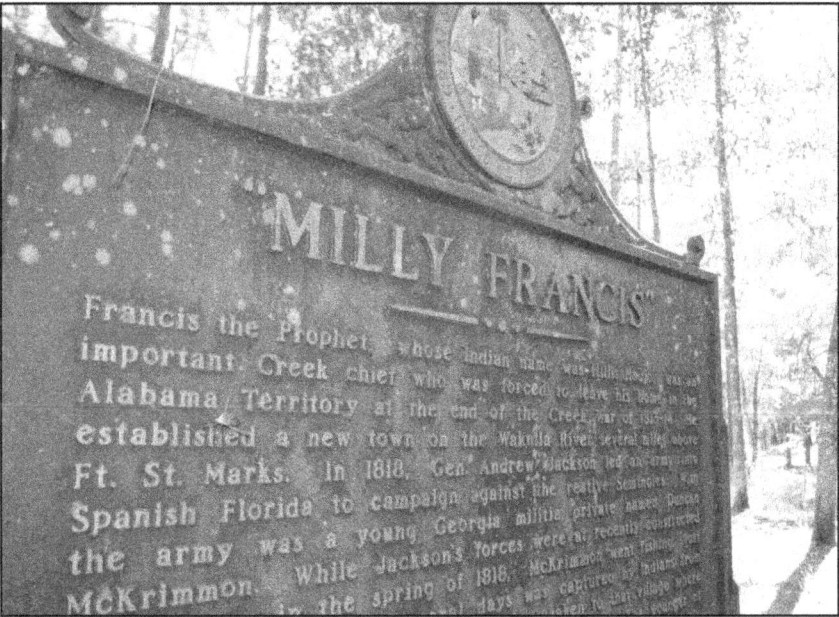

Historical marker for Milly Francis at Fort Gadsden (now at San Marcos de Apalache)

Capture of Homathlemico and Prophet Francis as pictured in 19th century

Scene of capture of Homathlemico and Josiah Francis

Photographs

Spanish ruins at San Marcos de Apalache Historic State Park

Ruins of Fort St. Marks (San Marcos de Apalache)

Brig. Gen. William McIntosh

Osceola as sketched from life by Catlin (Library of Congress)

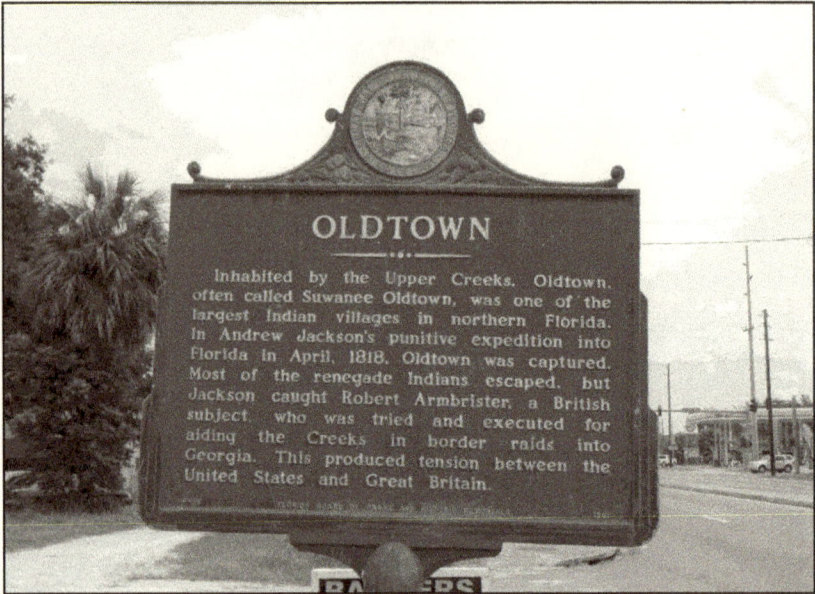

Historical Marker near site of Boleck's Town at Oldtown, Florida

Suwannee River near site of Boleck's Town

Chapter Thirteen

Repercussions

McIntosh and his warriors moved forward and joined Andrew Jackson's main column after clearing the battlefield of the prisoners, wounded and captured supplies. They found, according to staff officer with the Georgia militia, "many articles of soldier's clothing." These were believed to be part of the uniform supplies of the 4th Infantry that had been taken from Lieutenant Scott's boat. There was little doubt in the minds of the commanders and soldiers of the army that they had come up with at least some of the Red Sticks involved in the attack on the upper Apalachicola.[1]

Jackson and McIntosh were disappointed that Peter McQueen had escaped them. Along with Josiah Francis, he had been one of the key players in the Creek War of 1813-1814 and both American generals had hoped to come up with him for a long time. Interviewing some of the prisoners taken by McIntosh's warriors, General Jackson made one of them an unexpected offer:

...Upon the application of an old woman of the prisoners, I agreed that if McQueen was tied and carried to the commandant of St. Marks, her people should be received in peace, carried to the upper tribes of the Creek nation, and there provisioned until they could raise their own crops. She appeared much pleased with these terms, and I set her at liberty, with written conditions to the commandant of St. Marks to that effect. Having received no further intelligence from McQueen, I am induced to believe the old woman has complied with her part of the obligation.[2]

McQueen, of course, was never turned over to the American troops by his people. He led the survivors of his band across the Suwannee River and deeper into the Florida peninsula. It was there he died over the next several years from natural causes and not the vengeance of the U.S. Army.

The Battle of Econfina was far from the end of Andrew Jackson's Florida campaign. Pushing forward, he attacked Boleck's town on the evening of April 16, 1818. Adopting the strategy pursued by Cappachimico at Miccosukee, the Alachua chief fought a delaying action while his women and children escaped across the Suwannee River. General Gaines was ordered over after the fleeing inhabitants of both Boleck's village and the neighboring African settlement, but no more than a few people could be captured. The army was more successful, however, in capturing one of the two Englishmen that Jackson believed were responsible for instigating the war. Robert C. Ambrister was taken on the Suwannee, as were Peter Cook and John Arbuthnot (Alexander's son).[3]

After destroying the villages clustered around today's community of Oldtown, the army divided. The Georgia militiamen started for home via a cross-country march back to that state. McIntosh and his warriors began their homeward march by a separate route a short time later. The regulars and Tennessee horsemen, meanwhile, accompanied Jackson back to St. Marks. Arbuthnot's schooner, captured by Lieutenant Gadsden near the mouth of the Suwannee, was used to transport the army's sick and wounded to Fort St. Marks.[4]

A controversy developed at the fort shortly after its capture by the American troops. As the officers and soldiers were examining the Spanish soldiers and civilians taken there, two were found to be wearing clothes thought to have come from the men of Lieutenant Scott's command:

...After Fort St. Marks was occupied by the American troops a black man and Spanish soldier was reported to me as having been arrested clad in the American

uniform, recognised as part of the clothing of the fourth and seventh regiments, captured in the boat commanded by Lieutenant Scott, in ascending the Appalachicola river.

In explanation the Spanish commandant observed, that his soldiers and the Seminole Indians were in the habit of trading with each other, and that his negro, with others of his garrison, had received his permission to purchase some clothing reported to have been brought in by the Indians.[5]

This explanation by Captain Francisco Casa y Luengo did not sit well with the American officers. They viewed the purchase of supplies taken from Lieutenant Scott's boat by the Spanish soldiers as part of a growing body of evidence that the Seminole/Red Stick alliance had been supported by the garrison. The testimony of William Hambly and Edmund Doyle that Cappachimico, Francis, Homathlemico and other key Indian leaders had enjoyed free access to the fort and regular conferences with its commandant only solidified General Jackson's already firm belief that the war had been instigated by outside influences.

Already convinced of their guilt, the general ordered Robert Ambrister and Alexander Arbuthnot tried before a military court. The latter man was charged with:

1. Exciting and stirring up the Creek Indians to war against the United States and her citizens; he, A. Arbuthnot, being a subject of Great Britain, with whom the United States are at peace.

2. Acting as a spy, and aiding, abetting, and comforting the enemy, supplying them with the means of war.

3. Exciting the Indians to murder and destroy William Hambly and Edmund Doyle, and causing their arrest, with a view to their condemnation to death; and the seizure of their property, on account of their active and zealous exertions to maintain peace between Spain and the United States and the Indians, they being citizens of the Spanish Government.[6]

Ambrister, meanwhile, was charged with:

1. Aiding, abetting, and comforting the enemy, supplying them with means of war, he being a subject of Great Britain, at peace with the United States, and lately an officer of the British colonial marines.

2. Leading and commanding the Lower Creeks in carrying on a war against the United States.[7]

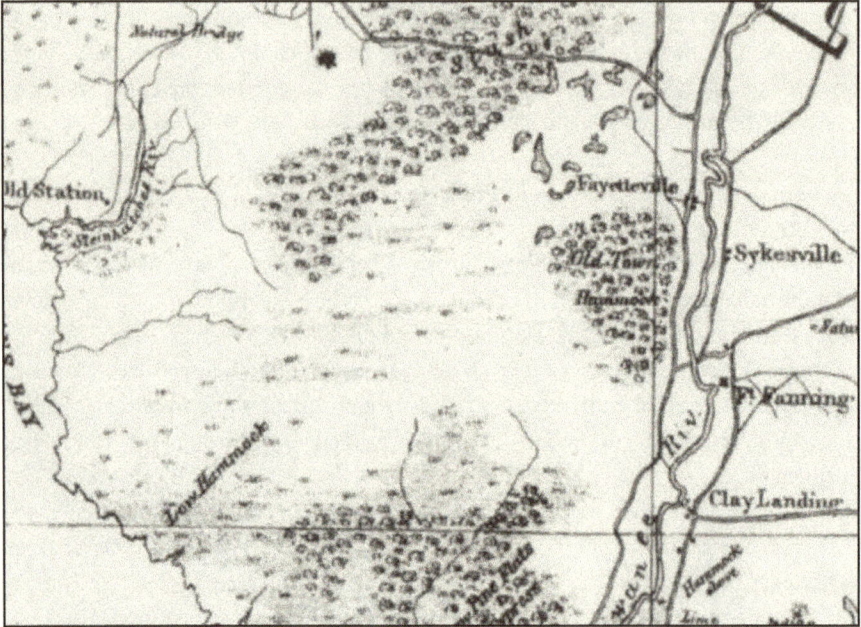

Map showing "Old Town Hammock," near the sites of the towns of Boleck (Bowlegs) and the Black Seminoles. The modern community of Oldtown stands in the vicinity.

The trials were quick and the sentences harsh. Both men were convicted and both men were executed. Arbuthnot was hanged from the yardarm of his own schooner, while Ambrister was shot by a firing squad. Their bodies joined those of Homathlemico and Josiah Francis in the marshy soil outside the walls of Fort St. Marks. The exact burial site has been lost to time.

The evidence against Ambrister proved far more compelling than that against Arbuthnot. The latter individual admitted providing supplies to the Indians but pointed out that his occupation was that of a trader. He did write numerous letters on behalf of the chiefs, representing them in efforts ranging from the attempt by Peter McQueen to secure the return of his slaves to the desperate pleas of Francis and Cappachimico for promised support from the Bahamas after the war was underway. Arbuthnot did send a letter of warning to his son that the American army was on the march, but overall does not seem to have taken an active role in the military side of the war effort. Ambrister, on the other hand, was demonstrated by his own writings and other evidence to have been what today would be described as "an enemy combatant out of uniform." He ordered out war parties to oppose the Americans, occasionally paraded in his British uniform coat and provided ammunition to the Indians. The testimony presented at his trial also offered compelling evidence that he had engaged in piracy by seizing a schooner off the mouth of the Suwannee River and holding its captain and crew as prisoners.[8]

The executions of the two British citizens would lead to an international outcry and strong exchanges of words between the United States and Great Britain, but neither country had the will to go to war again so soon after the close of the long and expensive War of 1812.

The First Seminole War continued for several more months. Andrew Jackson crossed the Florida Panhandle and seized Pensacola in May, while companies of rangers continued to battle individual bands of Red Stick and Seminole warriors until the following winter. U.S. troops held Fort St. Marks and Fort Gadsden in Spanish territory until the cession of Florida from Spain to the United States in 1821.

There was never more than minimal doubt after Jackson's successful invasion that Florida would become part of the United States. The Monroe Administration used the First Seminole War to present a case that Spain had failed to prevent the use of its Florida colony as a base for attacks on the United States (even though U.S. troops had instigated the war by attacking Fowltown north of the international border). Secretary of State John Quincy Adams placed the European nation on notice that it must put a sufficient military force in Florida to prevent a repeat of such attacks or the United States would be forced to seize the colony as an act of self-defense. The result was the signing of the Adams-Onis Treaty between the two countries in 1819. Spain agreed to surrender Florida to the United States, while the Americans gave up their claim that Texas had been included in the Louisiana Purchase. The Spanish colony of Florida became the U.S. Territory of Florida in 1821.

The transfer of Florida from Spain to the United States would not have taken place, at least in the way and at the time it did, had Seminole and Red Stick warriors not responded to the Fowltown raids by annihilating the command of Lieutenant Richard W. Scott. The attack was based from Spanish territory and carried out in Spanish territory. This fact may have been of little consequence to the Indians who felt they were responding to an unprovoked attack, but for the United States it provided an opportunity to change history. Jackson furthered this possibility by exceeding his orders and capturing both St. Marks and Pensacola. The Monroe Administration was quick to seize upon the situation and force Spain to negotiate the long desired transfer of Florida to the United States.

The state today is one of the five largest in the United States and at the time of this writing was home to more than 19.8 million people. It still revels in its Spanish heritage and Caribbean ties, but became part of the American nation nearly two centuries ago because of an all but forgotten battle on the banks of the Apalachicola River.

---

[1] Report dated Milledgeville, Georgia, May 5, 1818, citing an officer direct from the army, published in the Weekly Aurora, May 25, 1818, p. 111.

[2] Maj. Gen. Andrew Jackson to John C. Calhoun, Secretary of War, April 20, 1818, *American State Papers*, Military Affairs, Volume 1, pp. 700-701.

[3] *Ibid.*

[4] *Ibid.*; Maj. Gen. Andrew Jackson to John C. Calhoun, Secretary of War, April 20, 1818 (second report of this date), *American State Papers*, Military Affairs, Volume 1, p. 701.

[5] Lt. James Gadsden to Maj. Gen. Andrew Jackson, May 3, 1818, *American State Papers*, Military Affairs, Volume 1, p. 715.

[6] Minutes of the Trial of Alexander Arbuthnot, April 26, 1818, *American State Papers*, Military Affairs, Volume 1, p. 721.

[7] Minutes of the Trial of Robert C. Ambrister, April 27, 1818, *American State Papers*, Military Affairs, Volume 1, p. 731.

[8] Minutes of the Trials of Alexander Arbuthnot and Robert C. Ambrister, April 26 & 27, 1818, *American State Papers*, Military Affairs, Volume 1, pp. 721-735.

Chapter Fourteen

John & Elizabeth Dill of Fort Gaines

HOW AND WHEN ELIZABETH STEWART REACHED GEORGIA is not clear. She was returned to her husband and father by Brigadier General William McIntosh's staff following the Battle of Econfina. Whether she remained with Jackson's main army or returned north with the Georgia militia is not known. Her husband was a soldier of the First Brigade but few details have been uncovered about him other than that he died within a few years of her release from captivity at the Battle of Econfina.

The marriage records in Early County, Georgia, show that Elizabeth remarried John Dill on September 25, 1821. The present Clay County was then part of Early County and John Dill was a well-known merchant in the town of Fort Gaines, which grew up around the log fort of the same name. The fact that she remarried in Fort Gaines just two years after her release from captivity indicates that if Elizabeth ever left the frontier of Southwest Georgia, it was not for long.[1]

Tradition in Fort Gaines holds that General Dill, as he is remembered there, was an officer on the staff of Major General Edmund P. Gaines. Actual military records, however, tell a different story. Dill was not in the U.S. Army at the time

135

of the First Seminole War and was not on the staff of General Gaines. He had served as a private for four months during the War of 1812, but in the South Carolina militia and not the regular army. According to his state service records, John Dill was a private in Captain Turner's Company D, 3rd South Carolina Militia, from November 6, 1814, until March 12, 1815. He saw no action but was reported sick at the time of his discharge. He served a total of four months and 24 days and was paid $38 for his service.[2]

Within a few years of the end of his War of 1812 service, Dill showed up on the frontier at Fort Gaines where he opened a mercantile business. A thriving settlement was growing up around the fort, which was held by a small garrison of regular soldiers until 1821. By the time the troops headed west for Fort Smith, Arkansas, Fort Gaines had become a small town and important river port for a thriving agricultural region. What may have been the first house lived in by John and Elizabeth Dill still stands near the edge of the bluff at Fort Gaines. By today's standards it is a simple white wood-frame structure, but by the standards of 1821 it was impressive. Now preserved, it overlooks the site of the original fort as well as the earthworks and cannon of a battery constructed by the Confederates during the War Between the States.

By around 1827, however, Mr. and Mrs. Dill built a second and much more magnificent home. Now known as the "John Dill House," it stands on Washington Street near its intersection of Commerce Street in Fort Gaines. It was here the couple spent the most productive years of their lives.

Legend in Fort Gaines holds that while she was in captivity with the Red Sticks, Mrs. Stewart made a fortune through her own determination and ingenuity. The warriors of the various bands were then predatory raids against homes along the frontier and often came back to the camps in Florida loaded down with plunder. The horses and cows they brought back had obvious value, as did clothing, food and gold or silver coin. The paper money taken in such raids, however, was of no value to the Indians and according to the story they often threw it away onto the ground.[3]

Mrs. Dill (then Mrs. Stewart), of course, recognized the value of the paper being tossed away by the raiding parties and began to collect it and secret it away. By the time she was rescued by Major Kinnard at the Battle of Econfina, she had accumulated a small fortune. The money, it is said, helped the couple establish a mercantile empire in Fort Gaines and build two homes there, both of which still can be seen today.[4]

Whether the story is true or not, the Dills did accumulate wealth far above that of most of their neighbors. As their wealth grew, so too did his prominence in the region. By 1830, for example, he had been elected to the rank of major in the local militia regiment and was a key figure in the committee selected to organize that year's 4th of July celebration in Fort Gaines. His fellow committee members elected him as their chairman and he began the festivities with a toast to President Thomas Jefferson, who had died four years earlier. "He was a great and good man," Dill spoke aloud to the group, "his principals, the American political text book."[5]

One of the other toasts that day drew widespread approval when it celebrated the signing of the Indian Removal Act of 1830 by President Andrew Jackson:

...The Indian Bill; we hail it as a triumph of justice in favor of our slandered and oppressed State, and as a galling proof to those damnable Heralds, and their supporters, the National Journal, New York Daily Advertiser, New York Commercial Advertister, New England Palladium, Harrisburg Intelligencer, and Boston Patriot, with their satellites. Envy, hatred, outrage, devastation, and destruction, fiends like with a daring impudence stalking over the fairest portions of our land, struggling with a death like determination to wither and utterly annihilate the laureled wreath that binds our Federal Union. That Georgia in the majesty of her rights will be heard, the shafts of their calumny fall harmless at her shrine.[6]

The newspapers named in the toast, of course, had opposed the Indian Removal Act, a piece of legislation that drew wide favor in Georgia, Florida and Alabama. It authorized President Jackson to negotiate with Indian nations for their removal to new lands west of the Mississippi. The original intent of the act was for the Creek, Choctaw, Seminole, Cherokee and Chickasaw nations to be given the option of either remaining where they were and living under the laws of the individual states, or of relocating to new lands in what is now Oklahoma where they could retain some self-rule. The actual result, of course, was widespread violence and the forced removal of tens of thousands of people west on what became known as the Trail of Tears.

At the time of the 4th of July celebration in1830, Fort Gaines was just south of and across the Chattahoochee River from what remained of the Creek Nation. The First Seminole War had ended only twelve years before and the animosity that existed between the frontier settlers and the Creeks, particularly the Hitchiti and Yuchi who lived along the Chattahoochee River, was marked and intense. Men like John Dill, whose wife had spent five months in captivity, were fiercely dedicated to the removal of the Creek Indians from their traditional homes to new lands in the west.

A milestone was marked in Elizabeth's life in 1832 when she gave birth to a son, John P. Dill. By that point she was comfortably established in a large and beautiful home and her husband was a prominent and successful businessman and civic leader. The brutality she had seen on November 30, 1817, must have seemed far removed, but it was not long before the memories came surging back.

Major Dill was an ally and associate of William Wellborn, who soon played a key role in enforcing the removal of the Creeks from Alabama and who led troops against them in their last major fight with the whites at the Battle of Hobdy's Bridge in 1837. On January 10, 1834, Wellborn published the Executive Order from Milledgeville that promoted 26 men to the rank of colonel in the Georgia State Militia and designated them to serve as Aides-de-Camp to the state's commander in chief, Governor Wilson Lumpkin. John Dill was among them.[7]

Reappointed to the same role in the administration of Governor William Schley on January 13, 1836, Dill served as a colonel of the second brigade and aide-de-camp to the governor during the Creek War of 1836. Fort Gaines was on the very front of that war and, ironically, a portion of the warriors fighting against the United States was led by Neamathla himself.[8]

It was in his capacity as colonel, Dill joined two other officers in signing a letter to Governor William Schley in which they appealed for help to deal with the inability of the citizens of Fort Gaines to defend themselves against expected Indian attack:

...Rumors daily reach us of parties of these Indians, sometimes as high as seventy or [eighty] in company, having murdered in different parts of the country east of the Chatahoochee; and from various indications we have but too much reason to apprehend that the issues recently witnessed in East Florida, will be reenacted here. Our object in writing this letter is to apprise your Excellency that we are almost entirely destitute of arms, for our defence, having to depend exclusively upon such as individuals may happen to possess, and which would be by no means effective in an Indian war.[9]

How Mrs. Dill must have viewed such rumors just nineteen years after she had somehow survived the bloody destruction of Lieutenant Scott's party can be imagined. She must have remembered the scenes enacted before her eyes in 1817, scenes that she certainly hoped would never be enacted again.

The Creek War erupted in the spring of 1836 when an alliance of Yuchi and Hitchiti groups, led by Jim Henry, Neamathla and other chiefs, rebelled against efforts to force them from the last remaining sliver of Creek territory. Henry (who later became a Methodist minister in what is now Oklahoma under the name James McHenry) led a devastating raid against the town of Roanoke in Stewart County, Georgia, burning its structures to the ground and killing most of the handful of men left behind to defend the community when its women and children had fled to the safety of nearby Lumpkin and most of its men were visiting their families there.

The raid on Roanoke, which stood just up the Chattahoochee from Fort Gaines, prompted great alarm in the latter place. The citizens requisitioned any heavy timbers they could find and, as was noted in a letter to Governor Schley, built a fort for their own defense:

...We have built us a temporary Fort here, in doing which, we had to press all the Scantling & plank, Sills, & house framings, for it was built in a hurry, hearing that the Indians was on the way down from Roanoke., about ¼ of this Lumber is spoilt in Sawing, Short, & Cutting port holes, - We expect to be paid for the same, through the state from the Genl. Government.[10]

One day later, Colonel Dill appealed to Governor Schley to countermand orders requiring the militia companies from Fort Gaines and Early County to assemble with the main Georgia army at Fort Twiggs near Cusseta. The town, he reported, had become a point of last refuge for the citizens and the troops were needed there to protect them:

...[I]f any needs protection it is us. The point of the Creek Nation is only about four miles distant from Fort Gaines – and the white inhabitants of the nation has fled from the nation entire except Irwinton [i.e. Eufaula, Alabama], which is not yet abandoned by the inhabitants and which is daily expected to suffer the same fate of Roanoke, if so, Fort Gaines next, perhaps before, for we are in a defenceless situation at the present.[11]

Dill pointed out that the community of Fort Gaines was "the refuge for the women and children of this section." Included among those civilians he sought to protect, of course, was Elizabeth. The news from Roanoke had electrified the community and near panic ensued as citizens from throughout the area flooded into the town with what belongings they could bring.

By June 13, 1836, the colonel had defied orders and detained 18 drafted men from Early County for the defense of Fort Gaines. More than 100 women and children had evacuated to the town from Irwinton (Eufaula), Alabama, and the houses and buildings were crowded with suffering civilians. Exceeding his authority, Dill had formed a force to protect the hundreds of people gathered in Fort Gaines and prevailed upon the governor to approve of his actions. By raising two additional companies to augment the Fort Gaines Guards, he reported that he had been able to "keep a guard out at night which is of great relief to our women."[12]

Fighting did take place north, west and east of Fort Gaines, but the town itself was never attacked by the Creeks. Even so, May and June of 1836 must have been a terrifying time for all of the civilians assembled there, especially Mrs. Elizabeth Stewart Dill. Fortunately for Elizabeth, though, her life held no repeat of memories from the First Seminole War.

The Creek Indians were forcefully removed to what is now Oklahoma and after the last major battle at Hobdy's Bridge in 1837, any fear of possible Indian attack on Fort Gaines faded away. John Dill served in the Georgia militia for more than another decade and in 1847 was reported to be the brigadier general of the 1st Brigade. This service was the origin of his title of "general," by which he is often referred today.[13]

General Dill lived until the 1850s and spent the rest of his life as a prominent and successful businessman who owned mercantile establishments, a brickyard and a cotton warehouse. After he passed away, Elizabeth applied for and received a pension based on his War of 1812 service. She collected $3.50 per month for the rest of her life. By 1860, Elizabeth Dill was living in the home of James and Sarah Touson (Towson). Her personal wealth was reported by the census taker to be in the range of $5,000, an impressive sum for a 68-year-old widow who had lived much of her life on the frontier. She died in Fort Gaines on September 5, 1864, and was buried in the town's Old Pioneer Cemetery alongside her husband.[14]

---

[1] Jordan Dodd, *Georgia Marriages to 1850* [database on-line], Provo, UT, USA: Ancestry.com Operations Inc, 1997.

[2] U.S. Army Register of Enlistments, 1798- ; Service Record of Private John Dill, State Militia Records, South Carolina State Archives.

[3] Traditional story still told in Fort Gaines.

---

[4] *Ibid.*

[5] "4th of July at Fort Gaines, Geo.," *Georgia Journal*, July 24, 1830, p. 3.

[6] *Ibid.*

[7] Executive Order dated Milledgeville, January 10, 1834, published in the Macon *Weekly Telegraph*, January 23, 1834, p. 3.

[8] Executive Order dated Milledgeville, January 13, 1836, published in the Macon *Weekly Telegraph*, January 28, 1836, p. 3.

[9] J. Patterson, A. McGinty and J. Dill to Gov. William Schley, January 29, 1836.

[10] William P. Ford to Gov. William Schley, May 26, 1836.

[11] Col. John Dill to Gov. William Schley, May 27, 1836.

[12] Col. John Dill to Gov. William Schley, June 13, 1836.

[13] Eileen Babb McAdams, "Georgia Militia 1847, Major and Brigadier Generals," (List of officers assembled from *Augusta Chronicle*, March 31, 1848), U.S. Genweb.

[14] U.S. Census for District 749, Clay County, Georgia, 1860.

Chapter Fifteen

Fates and Futures

ANDREW JACKSON'S SUCCESSFUL INVASION and the cession of Florida from Spain to the United States three years later forever ended the hold of the Seminoles and Lower Creeks on Northwest Florida. Individual groups remained and some hostilities took place in the region during the Second Seminole War, but the Treaty of Moultrie Creek had wiped away Indian claims of ownership to the vast region. The Seminoles and Red Sticks were forced south down the peninsula and many of the Lower Creeks were required to move up into the main Creek Nation. The future of the frontier had been altered forever by the outcry that followed the Scott Battle of 1817. The road to the Trail of Tears had been paved.

Major David E. Twiggs, the U.S. Army officer in command at Fort Scott when the confrontation developed with Neamathla and the man who led the first attack on Fowltown, remained a figure on the frontier for the rest of his life. Promoted to lieutenant colonel in 1831 and colonel in 1836, he served with distinction at the Battles of Palo Alto and Resaca de la Palma during the Mexican War. He was promoted to brigadier general and given the brevet rank of major general for gallantry during that war. Congress was so appreciative of his service that it voted him a sword and gold scabbard.

Twiggs assumed command of the Department of Texas in 1857 and was Colonel Robert E. Lee's commanding officer when the Southern states began to secede from the Union in 1860. General Twiggs repeatedly warned Washington that he would not carry on a civil war against the people of Texas and determined that he would never open fire on his fellow citizens. At the same time, however, he refused to require his men to surrender up their arms, a point over which he told commissioners from the State of Texas he was willing to give his life.

In the end, surrounded by Texas volunteers and under guard, General Twiggs surrendered all public property and forts in Texas on the condition that his soldiers be allowed to march to the coast with their small arms. Dismissed from the U.S. Army by order of President James Buchanan, the old hero of the War of 1812, Seminole and Mexican Wars accepted the rank of senior major general and command of the District of Louisiana for the Confederacy. General Twiggs was already 71 years old, however, and his health entered serious decline. He returned home to Richmond County, Georgia, and he died there on July 15, 1862.

Neamathla remained in Florida after the war, settling down near the old Tallahassee Talofa site in what became Leon County. He joined with John Blunt and Econchattimico in visiting Andrew Jackson at Pensacola in 1821 to discuss what they and their people should do now that Florida had become a U.S. possession. Jackson advised them to return to their homes and remain at peace. In fact, the general turned governor even pleaded their case to Washington, advising that they be assigned lands and allowed to remain where they were so long as they desired.

The chief and his people were still living at Tallahassee when commissioners John Lee Williams and Henry Simmons arrived there looking for a site to plan a capital for the new Territory of Florida. They met with Neamathla, who was unhappy with the prospect and who warned the commissioners to remain silent about their plan as he might not be able to restrain his other chiefs and warriors. He subsequently engaged in something of a war of words with incoming governor William P. Duval. The governor went to Neamathla's town to stand down an alleged plan by the chief to attack Tallahassee's earliest settlers, a confrontation that was immortalized by famed American writer Washington Irving in "Conspiracy of Neamathla." The short story appears in the book *Wolfert's Roost* which Irving published in 1855 and was based on the personal account of Governor Duval.

Neamathla was among the chiefs who assembled at Moultrie Creek in 1823 and it was there that he delivered one of the most impassioned speeches ever given by a North American chief:

We are...poor and needy; we do not come here to murmur or complain; we want advice and assistance; we rely upon your justice and humanity; we hope that you will not send us south, to a country where neither the hickory nut, the acorn, nor the persimmon grows; we depend much upon these productions of the forest for food: in the south they are not to be found. For me, I am old and poor; too poor to move from my village to the south. My cabins have been built with my own hands; my fields cultivated by only myself. I am attached to the spot improved by my own labor, and cannot believe that my friends will drive me from it.[1]

Concluded on September 18, 1823, the Treaty of Moultrie Creek assigned to Neamathla a reservation of four square miles in eastern Gadsden County. The chief never occupied it, but instead moved with his people to the Creek Nation in Alabama. He settled among the Yuchi and Hitchiti in an area just north of present-day Eufaula and was living there thirteen years later when part of the Nation rose up against the whites in the Creek War of 1836. A prominent leader in that war, Neamathla was over 80 years old when he was captured by Alabama troops, locked in irons and forced west on the Trail of Tears. It was reported by army officers who saw him that he never voiced a complaint.

The old chief survived the long journey west to what is now Oklahoma where he lived out his final years and died in the Creek or Muskogee Nation. The final mentions of him that appear in government records are complaints he lodged against the U.S. government for not supplying the blankets and other supplies that his people had been promised and so desperately needed.

Major General Edmund Pendleton Gaines continued to serve on the frontier following the close of the First Seminole War and by 1836 was in command of the Western Military Department. When he learned of the December 1835 massacre of Major Francis Dade's command by Seminoles in Florida, he immediately went into action and moved a strike force to Florida as rapidly as possible without waiting to hear from the War Department. This later embroiled him in a controversy with Major General Winfield T. Scott, who had been assigned overall command of U.S. troops in the Second Seminole War.

Gaines and his command was the first to visit the scene of the Dade Battle and it was under his orders that the slain soldiers still littering the ground were identified and buried. He attempted to advance against the Seminole stronghold in the cove of the Withlacoochee River with more than 1,000 men, but was fought to a standstill and held under siege at the hastily constructed Camp Izard. It was during this battle that Gaines received a bullet wound that knocked out several of his teeth. He supposedly spat them out on the ground and remarked that it was mean of the Seminoles to cost him his teeth when he had so few left.

Returning to the Texas frontier that same year, Gaines watched as General Antonio Lopez de Santa Anna advanced north across Texas driving everything before him in what became known as the "Running Scrape." He positioned his forces along the Trinity River to await developments and defend the United States should Santa Anna attempt an invasion of Louisiana in retaliation for the Texas Revolution. The Mexican general's army was smashed at the Battle of San Jacinto by the forces of General Sam Houston, however, and Gaines was never forced to lead his troops into action. His presence and efforts to prevent the nearby Cherokee and Caddo Indians from involving themselves against the Texans, however, were credited for enabling Houston to succeed at San Jacinto.

The general served in the Mexican War and was court martialed for calling up volunteer forces to support General Zachary Taylor's army without first securing the permission of Washington. He was acquitted of the charges and remained a key U.S. Army officer until he died of cholera in New Orleans on June 6, 1849. He was 72 years old at the time and was considered an American hero, especially for his gallantry during the War of 1812.

Cappachimico, the chief who had so eloquently responded to the demand from General Gaines for the surrender of the murderers of Mrs. Garrett and her family, survived the Battle of Miccosukee. Although General Jackson initially believed that the old chief had been killed, he actually led his people to safety while fighting a rearguard action as the army attacked his towns. When the troops departed, he helped re-establish Miccosukee at a new site near Greenville in what is now Jefferson County, Florida. He died there shortly thereafter and by 1821 and was replaced by John Hicks (Tuckose Emathla), a chief who also had fought against Andrew Jackson in 1818.

Lieutenant Colonel Matthew Arbuckle was promoted to full colonel in 1820 and given command of the 7th U.S. Infantry. He personally directed the movement of the troops from Fort Scott west to Fort Smith, Arkansas, in 1821 and in 1824 established Fort Gibson in what is now Oklahoma. He commanded along the western frontier for many years. He assumed command again at Fort Smith in

1848 and the following year assigned troops to provide security for prospectors heading west in the California Gold Rush. He died of cholera on June 11, 1851 and is buried on Arbuckle Island near Fort Smith. The Arbuckle Mountains of Oklahoma are named in his honor.

Milly Francis, the 15-year-old daughter of the Prophet Josiah Francis, was among the followers of the Prophet who surrendered themselves at Fort St. Marks following his execution. They were given supplies and directed from there to Fort Gadsden to begin their return trip to the Creek Nation.

The story of her intervention to save the life of Private Duncan McCrimmon (McKrimmon) of the Georgia Militia inspired the passions of newspaper readers across the nation and donations were sent south to Colonel Arbuckle at Fort Gadsden with requests that they be given to her. McCrimmon also returned to the Apalachicola to propose marriage to Milly in gratitude for her having saving his life, but she refused his offer and chose to remain with her people instead.

The young woman eventually married a man of her own nation, but he died while fighting for the United States in the Second Seminole War. She was among the Creek men, women and children who were rounded up and placed in concentration camps during and following the Creek War of 1836. From there she took part in the long journey west on the Trail of Tears to a new home in what is now Oklahoma.

Milly reached Oklahoma in January 1838 and settled with her children in a dirt-floored log cabin on or near the grounds of what is now Bacone College in Muskogee, Oklahoma. Lieutenant Colonel Ethan Allen Hitchcock visited her there a few years later and found that she was living in abject poverty. Moved by her condition and her selfless act in saving the life of McCrimmon, he wrote to the Secretary of War urging that something be done to help her. The Secretary presented the idea to the U.S. Congress and in 1844 Milly Francis was voted a pension of $96. Congress also approved the casting of a medal honoring "Milly, a Creek woman." Although the Congressional Medal of Honor did not yet exist, Milly could rightfully be said to be the first woman ever awarded a special medal of honor by Congress.

Milly Francis died of tuberculosis in 1844 and was believed to have been buried near her home. She is honored today by a stone monument on the campus of Bacone College, as well as a smaller monument at San Marcos de Apalache Historic State Park in St. Marks, Florida. She is often referred to as the "Creek Pocahontas."

Major General Andrew Jackson continued to serve in the U.S. Army until 1821 when he resigned to accept the appointment of President James Monroe as the first governor of the Territory of Florida. He held the position only until December of that year before returning home to the Hermitage in Nashville, Tennessee.

A powerful figure and national hero due to his smashing victory over the British at the Battle of New Orleans, Jackson and his allies began a slow but steady march to the White House. Elected a U.S. Senator from Tennessee, Jackson waged a fierce campaign for the Presidency in 1824. He received the largest popular vote of the four candidates, but the U.S. House of Representatives awarded the office to John Quincy Adams. Jackson and his followers felt the election had been stolen from them.

Returning with a vengeance in the 1828 campaign, Jackson swept the popular vote and rode a wave of support into the White House. A populist who hated the power exerted over the affairs of the nation by the wealthy elite of the Northeast, the new President set about breaking their grip on the country. He dismantled the 2nd National Bank of the United States, balanced the federal budget, paid off the national debt (the only President ever to do so) and forced South Carolina to back down from threats to secede during the Nullification Crisis, a dispute over whether a state could nullify or ignore a federal law with which it disagreed.

President Jackson was a supporter of and signed the Indian Removal Act and is given much of the blame today for the Trail of Tears and the deplorable way in which the people of the so-called Five Civilized Tribes (Creek, Cherokee, Seminole, Choctaw and Chickasaw) were forced from their lands and removed to what is now Oklahoma. Over 4,000 Cherokee and thousands of others died during the forced emigration.

Andrew Jackson was the first target of an attempted presidential assassination in U.S. history. As he was leaving a state funeral at the Capitol building, an unemployed English housepainter named Richard Lawrence approached the President and aimed a pistol at him but the weapon misfired. Lawrence then drew a second pistol, but it also misfired. Jackson attacked the man with his cane and beat him until a number of other officials, including Rep. David Crockett of Tennessee, restrained Lawrence and took him away into custody. The would-be assassin later claimed to be the King of England. Remarkably, when the pistols used by Lawrence were test fired later, they performed perfectly each time. Their failure to work against Andrew Jackson was attributed to Providence.

Jackson served two terms as President, finally leaving the White House to return home to Tennessee in 1837. He remained active and lived out his years at the Hermitage, promising on his deathbed that he would see his friends – white and black – again in Heaven. "Old Hickory" died on June 8, 1845, of tuberculosis, heart failure and edema (Dropsy). He is buried on the grounds of his beloved home.

Osceola fled south from the Battle of Econfina with the survivors of Peter McQueen's band. His uncle led the desperate and starving men, women and children down in the Florida peninsula to an area near Paynes Prairie where they united with other Creeks and Seminoles who were clustering there. Although he had been born and raised as a Creek, he grew to be a warrior among the Seminoles.

In 1832, when he was around 28 years old, Osceola was among the chiefs and warriors who gathered at Payne's Landing for treaty negotiations with U.S. commissioners led by Indian Agent Wiley Thompson. The warrior was so infuriated by the proposed terms that he stabbed his dagger into the treaty document. Thompson had him confined in chains until he agreed to abide by the terms of the Treaty of Payne's Landing, which required the Seminole to move to new lands in Oklahoma.

Like Osceola, however, many of the chiefs alleged that they had been bullied at the treaty negotiations and tensions grew over the next several years. Things finally exploded in 1835 when a war party led by Osceola killed the accommodation-minded chief Charley Emathla and scattered the money he had received for agreeing to leave Florida. Then, on December 28, 1835, he led a war party that shot and killed Agent Thompson and several other men as they were walking outside Fort King at present-day Ocala. A much larger force attacked and destroyed the command of Major Francis Dade on the same day. The Second Seminole War had begun.

Osceola took part in the fighting of the next eighteen months, particularly in and around the Cove of the Withlacoochee. On October 21, 1837, he was decoyed to a truce negotiation at Fort Peyton near St. Augustine and seized by order of General Thomas Sidney Jesup. After a brief imprisonment at the Castillo de San Marcos (Fort Marion), Osceola was moved to Fort Moultrie in South Carolina where he died of malaria on January 30, 1838. His grave can be seen there today near the sally port of the old fort. He was 33 years old.

Osceola is remembered today as a symbol of the determined resistance of the Florida Seminoles. He is honored by Florida State University in Tallahassee where he is a symbol for the university in a tradition supported by the Seminole Tribe of Florida. Osceola National Forest in Florida is named in his honor.

Elizabeth Dill outlived most of her contemporaries, including Osceola, McQueen, Twiggs, Gaines, Jackson and even Milly Francis. As was noted in the previous chapter, she spent the rest of her life in Fort Gaines watching that community grow from a frontier outpost to a thriving commercial center. Two of her houses still stand there and the legend lives on that she funded them using paper money collected while a captive with the Red Sticks in Florida. She died in 1863 and is buried at the Old Pioneer Cemetery in Fort Gaines. Her capture at the Scott Battle of 1817 is recreated in a mural in Dothan, Alabama.

Within three years of the Scott Battle, Florida became part of the United States. It is now one of the five most populous states in the Union.

---

[1] Speech of Neamathla, September 11, 1823, *American State Papers*, Indian Affairs, Volume II, p. 439.

Chapter Sixteen

The Scott Battle in the News

The following accounts of the Scott Battle appeared in newspapers across the United States during the weeks and months following the battle.

December 26, 1817
New York Commercial Advertiser

War with the Indians. – An Officer in the army stationed at Fort Scott, under date of the 2d of December communicates to his Father in Baltimore, the following unpleasant information. – "Lieutenant Scott, of the seventh, had been ordered down the Appalachicola with about forty men, to assist the vessels which were coming up with supplies. Major Meckenburg, who commanded the troops on board the vessels, ordered Lieut. Scott back, and put on board the clothing of our regiment, and several women and children. Yesterday, five of his men came in, all wounded. They state, that Lieut. Scott was attacked by the Indians just below the forks of the rivers, and the whole party killed except themselves.

"This is truly lamentable. I expect we shall have some very warm work before many days. The whole Indian force is supposed to be 2800.

"P.S. – Since I commenced this letter, the Indians have fired upon some women who were washing on the bank."

December 31, 1817
Massachusetts Spy

Extract of a letter from an officer in the army, to his father in Baltimore, dated "Fort Scott, Dec. 2.
"I have just been informed, that an express will start for fort Hawkins in twenty minutes. I therefore send you this hasty note. I marched from fort Hawkins on the 15th Nov. and arrived here on the 19th at night. On the 22d, col. Arbuckle crossed Flint river with 300 men, for the purpose of destroying an Indian town about 20 miles off. We arrived in the town about 12 o'clock; next day, at 3, the Indians attacked us, and, after an action of about fifteen minutes, they retreated into a large swamp, which nearly surrounded their town. Their loss cannot be ascertained. Ours, one killed, one severely and three slightly wounded. Lieut. Scott, of the 7th, had been ordered down the Apalachicola, with about forty men, to assist the vessels which were coming up with supplies. Maj. Muhlenberg, who commanded the troops on board the vessels, ordered Lieut. Scott back, and put on board the clothing of our regiment, and several women and children. Yesterday five of his men came in all wounded. They state, that Lieut. Scott was attacked by the Indians just below the forks of the river, and the whole party killed except themselves. This is truly lamentable. I expect we shall have some very warm work before many days. The whole Indian force is supposed to be 2,800.
"P.S. Since I commenced this letter, the Indians have fired upon some women who were washing on the bank."

January 7, 1818
Vermont Rutland Herald

The *Savannah Republican* of the 17th ult. States that more lives were lost in the last attack upon Lieutenant Scott, by the Indians, than was at first supposed. It appears hat Lieutenant Scott, 44 men, 10 women and three children were killed making in all 58. The clothing for the 4th regiment, under the guard of Scott, was also taken off by the savages.

January 14, 1818
Independent American

Copy of a letter from Major General Edmund P. Gaines, to Governor Rabun, of Georgia, (received by express yesterday morning) dated "Head Quarters, Fort Scott, Dec. 2, 1817."

Sir – I have the honor to acknowledge the receipt of your excellency's letter of the 20th of last month. The detachment of militia, I have no doubt, will arrive in due time to enable to put an end to the little war in this quarter, in the course of this or the next month.

With a view to ascertain the strength of the hostile Indians in the vicinity of Fowl Town, and to reconnoiter the adjacent country, I, a few days past detached Lieut. Col. Arbuckle, with 300 men. The Lieut. Col. reports, that a party of Indians had placed themselves in a swamp, out of which about 60 warriors approached him, and with a war hoop commenced a brisk fire upon the detachment. They returned the fire in a spirited manner. It continued not more than 15 or 20 minutes before the Indians were silenced, and forced to retire into the swamp, with a loss which Lieutenant Col. Arbuckle estimates at 6 to 8 killed, and a much greater number wounded. We had one man killed, and two wounded. The enemy have since succeeded in an affair in which the real savage character has been fully exhibited. A large party formed an ambuscade on the 30th ultimo, upon the Appalachicola river, a mile below the junction of the Flint and Chattahoochie, attacked one of our detachments in a boat, ascending near shore, and killed, wounded and took the greater part of the detachment consisting of 40 men, commanded by Lieutenant R.W. Scott. There were also on board the boat, killed or taken, 7 women, the wives of soldiers; six men only escaped, four of whom were wounded. They report that the strength of the courrent at the point of attack, had obliged the Lieut. to keep his boat near the shore. That the Indians had formed along the banks of the river, and were not discovered until their fire commenced, in the first volley of which, Lieutenant Scott and his most active men fell. The Lieutenant and his party had been sent from this place some days before, to assist Major Muhlenberg in ascending the river with three vessels, laden with military supplies, brought from Fort Mongomery and Mobile. The Major, it seems, deemed it proper to retain only about 20 men of the party, and in their place put a like number of sick, with the women, and some regimental clothing. – The boat thus laden, was unfortunately detached alone for this place. It is due to Major Muhlenberg, to observe, that at the time he detached the boat, I have reason to believe he was not apprised of any recent acts of hostility having taken place in this quarter. It appears, however by a letter from Lieutenant Scott, received about the hour in which he was attacked, that he had been warned of the danger which awaited him. I must, therefore, conclude that he felt it to be his duty to proceed. Whether he had received from Major Muhlenberg a positive order to this effect, I have not yet learned. Upon the receipt of Lieutenant Scott's letter, I had two boats fitted up with covers of plank, port holes, &c. for defence, detached them under Captain Clinch, with a sub-altern officer and 40 men, with an order to secure the movement of Lieutenant Scott, and then to assist Major Muhlenberg. This detachment embarked late in the evening of the 30th ult. And must have placed the scene of the action (15 miles below this place) at night, and 7 hours after the affair had terminated. I have not yet heard from captain Clinch. I shall immediately strengthen the detachment under Major Muhlenberg with another boat secured against the enemy's fire. He will, therefore, move up safely by keeping near the middle of the river, with his vessels and force, is quite practicable. I shall moreover, take a position with my principal force, near the junction of the rivers at the line of demarcation between the United States and Spain, and shall attack any force near that place, or that may attempt to intercept our vessels or supplies below.

The wounded men who made their escape, concur in the opinion that they had seen upwards of 500 warriors (supposed to be hostile) at different places on the river, below the point of attack; of the force engaged they differ in opinion, but all agree that the number was very considerable, extending about one hundred and fifty yards along the shore, at the edge of a swamp in a thick wood.

I am assured by the friendly chief, that the hostile warriors of the town on the Chattahoochie, have been for some time past moving off down the river, to join the Seminoles. Those now remaining on that river, are believed to be well disposed. One of the new settlers there, however, has recently been killed there; but it has been already proven, that the perpetrator of this act, together with most of the warriors of this town (High Town) belonged to, and have joined the hostile party. The friendly chief in the neighborhood, promptly dispatched a party in pursuit of the offender, who made his escape towards the Mickasokee town. Onishays, and several other friendly chiefs, have tendered to me their services, with their warriors, against the Seminoles. I have promised to give them notice of the time that may be fixed on for my departure, and then to accept of their services.

The enclosed paper contains the substance of what I have said to the chiefs who have visited me; several of whom reside south of the Apalachicola.

The chiefs were desirous I should communicate to them my views and wishes. I felt authorised to say but little, and deemed it necessary in what I should say to counteract the erroneous impressions by which they have been misled by pretended British agents.

April 11, 1818
Camden Gazette

(Addressed to Major Daniel Hughes at Fort Mitchell)
Uche Old Fields, March 2.
Sir: - I wish you would inform our Agent and our head men, that since I left Fort Mitchell, the fourth day at twelve o'clock, I have taken three of our enemies that were firing on the vessels on this river, and one was wounded at the same place when firing on the vessels. I have got them in strings, carrying them to Fort Gaines, and expect to catch some more before I get there. Nothing more but the creeks are very high – it is as much as we can do to travel.
I remain your friend.
William M'Intosh,
Gen'l commanding.

Fort Gaines, March 6.
Major Daniel Hughes – I wrote you the other day and told you that I had taken 3 prisoners – I carried them to Fort Gaines to the commanding officer, and he told me he would have nothing to do with them, and said to me, you may deal with them by your own laws. We had proof that they were at the destroying of the boat below the fork of Flint river, and one of them was wounded at that time – they were doing mischief to our friend and I knew what was the law between us and the United States; I did not want them to stand on our land, and I have taken their lives – I have heard where a good many of our enemies are collected, about forty miles from this place, and I am going to push on to them to-morrow as fast as I can get where they are. This is all I have to say to you and our head men and agent, and whatever I do hereafter I will let you know again. Nothing more; all my men are healthy – your friend,
Gen. Wm. M'Intosh,
Com'g the Creek Indians.

May 16, 1818
Berks and Schuylkill Journal

(Addressed to D.B. Mitchell, Agent for Indian Affairs)

Camp, 3 miles from Mickasukie, (on the way to Sawanee,) 13th April, 1818.

SIR – Since I left you I have not sent you a talk of what we have done, and I now send you this. I heard yesterday of Peter McQueen being near the road we were travelling, and I took my warriors and went and fought him. There seemed to be a considerable number collected there. When we first began to fight them, they were in a bad swamp, and fought us there for about an hour, when they ran, and we followed them three miles. They fought us in all about three hours. We killed 37 of them, and took 98 women and children and six men prisoners, and about 700 head of cattle and a number of horses, with a good many hogs and some corn. We lost three killed and had five wounded. Our prisoners tell us that there was 120 warriors from six different towns. From what we saw I believe there was more than they say, as some of our prisoners say there was 200 of them. Tom Woodward* and Mr. Brown, and your Son, our Agent, and all the white men that live in our country, were with us through the whole fight, and fought well. All my officers fought so well I do not know which is the bravest. They all fought like men and run their enemies. General Jackson waited for us about six miles from where we fought. After the fight I went and joined him, and we are going this morning to fight the negroes together. They are at Suwannee, and we shall be there in four days. There was among the Hostiles a woman that was in the boat where our friends the white people were killed on the river below Fort Scott. We gave her to her friends – her husband and father are with general Jackson – Major Kinnard took her himself. This is all I have to tell you. I wish you would send a copy of this to the Big Warrior and Little Prince.
Your Friend
WILLIAM MCINTOSH,
Brig. Gen. Comm'g C.W.
--

*Major Woodward of Baldwin

(Milledgeville, (Geo.) April 28.
...McIntosh, in the late battle, killed with his own hand, three of the enemy – this fact is communicated by Mr. Wm. Mitchell (son of the Agent) in a letter to his father.

May 25, 1818
Weekly Aurora

Milledgville, May 5

Latest from the army. – For the following late intelligence from our troops, we are indebted to Mr. Pearre, (one of the editors of the *Augusta Chronicle*,) an officer of col. Milton's staff, who is direct from the army.

The army left St. Marks on the 9th ult. and on the 10th were joined by a detachment of mounted troops from Tennessee, under the command of colonels Dyer and Williamson, a small detachment of regulars under the command of captain Call, and 1000 warriors under McIntosh, who had been left at the Mickasukie to scour the country and gather the stock which was left by the Indians in their retreat. On the morning of the 12th, McIntosh with a part of his warriors, attacked a party of hostile Indians. The engagement continued about two hours with much spirit, when the hostiles retreated, leaving their women, children and property of all kinds to the mercy of the conquerors – Sustaining a loss of 37 killed on the field, and two wounded, and a number of prisoners. The number of Indians engaged was differently represented by different prisoners. McIntosh had three men killed, and several wounded. Kinnard and Timpoochee (or John) Barnett were conspicuous in this action; the latter evinced military talents which would have done credit to a greater man. These Indians belonged to McQueen's party, and were the same who massacred the crew of lieutenant Scott in the boats last fall, at the mouth of Flint river. The woman who was taken there, and many articles of soldier's clothing were found in their possession.

On the 17th the army took possession of Suwaney, after a skirmish of about fifteen minutes, in which three negroes were killed, and three taken prisoners. About 2000 bushels of corn, some cattle, and some few articles of provisions such as rice, potatoes, sugar, salt, &c. were found in the town, and at a store belonging to Arbuthan, a few miles below. On the next day a scout was sent across the river for the purpose of of pursuing the Indians, but they had got too far advanced to be overtaken. The scouts took some property, and found a small quantity of merchandize concealed in the swamp.

On the night of the 18th, two Englishmen who Arbuthnot had employed as clearks and agents, and two negroes, came from a schooner just arrived below from a piratical cruise, up to the town for provisions, &c. unconscious of our army being there. They were all taken by our centinels except one negro, who made his escape. The canoe which they came in was secured, and at day light next morning a detachment was sent to take possession of the schooner, on board of which young Arbuthnot commanded. The result of the expedition was not known when our informant departed.

On the 20th, the Georgia troops commenced their march homeward. In the evening of the same day, McIntosh and the principal part of his warriors also commenced their return march, with directions to destroy Hoponnie's town and all his warriors, and to take possession of all his property of every description, so as effectually to destroy him.

January 7, 1819
National Advocate

(Quoting a letter of the Secretary of State of the United States to the Minister Plenipotentiary of the U.S. to Spain at Madrid)

After the repeated expostulations, warnings and offers of peace, through the summer and autumn of 1817, on the part of the U. States, had been answered only by renwed outraged, and after a detachment of forty men, under Lieutenant scott, accompanied by seven women, had been waylaid and murdered by the Indians, orders were given to General Jackson, and an adequate force was placed at his disposal, to terminate the war.

January 22, 1819
Rhode Island America

Mr. Johnson, of Kentucky, also of the Military Committee, submitted a paper drawn up in the shape of a Report by that Committee, which, by a majority of one vote, that Committee had refused to accept, and the said paper was read as follows:

...Under this influence, they not only refused to deliver the murderers, but repeated their massacres whenever opportunity offered; and, to evade the arm of justice, took refuge across the line, in Florida. In this state of affairs, in November, 1817, Lieutenant Scott, of the United States' army, under General Gaines, with 47 persons, men, women and children, in a boat on the Appalachicola river, about a mile below the junction of the Flint and the Catahoochie, was surprised by an ambuscade of Indians, fired upon, and the whole detachment killed and taken by the Indians, except six men who escaped by flight (one of whom was wounded.) Those who were taken alive on this occasion were wantonly murdered by the ferocious savages, who took the little children, and dashed out their brains against the side of the boat, and butchered all the helpless females, except one, who was afterwards retaken....

Illustrations

Chapters Thirteen - Sixteen

First home of John and Elizabeth Stewart Dill in Fort Gaines, Georgia

Second home of John and Elizabeth Stewart Dill in Fort Gaines, Georgia

Cemetery where Elizabeth Stewart Dill is buried in Fort Gaines, Georgia

Final resting place of Elizabeth Stewart Dill in Fort Gaines, Georgia

Milly Francis Monument at Bacone College in Muskogee, Oklahoma

Milly Francis Monument at San Marcos de Apalache in St. Marks, Florida

James Monroe, U.S. President who authorized retaliation for the Scott Battle

John C. Calhoun, Secretary of War during the First Seminole War and later nemesis of President Andrew Jackson

John Quincy Adams, negotiator of the Adams-Onis Treaty with Spain

Casualties of the Scott Battle

November 30, 1817

<u>Killed by Company</u>

Name	*Rank*	*Unit*
Richard W. Scott	1st Lieutenant	Twiggs, 7th Infantry
Frederick McIntosh	Sergeant	Twiggs, 7th Infantry
James Edwards	Corporal	Twiggs, 7th Infantry
John Asbury	Private	Twiggs, 7th Infantry
David Brooks	Private	Twiggs, 7th Infantry
John Henderson	Private	Twiggs, 7th Infantry
William James	Private	Twiggs, 7th Infantry
Samuel McDonald	Private	Twiggs, 7th Infantry
Henry Moore	Private	Twiggs, 7th Infantry
Smith Irvin	Private	Twiggs, 7th Infantry
George Mullis	Private	Twiggs, 7th Infantry
James Thompson	Private	Allison's, 7th Infantry
John Gravit	Private	Bee's, 7th Infantry
Jesse Greenlee	Private	Bee's, 7th Infantry
Alfred Simmons	Private	Bee's, 7th Infantry
Henry Williams	Private	Bee's, 7th Infantry
William Darby	Private	Birch's - 7th Infantry
Jonathan Driver	Private	Birch's, 7th Infantry
James Holley	Private	Birch's, 7th Infantry
Charles Craft	Private	Corbaly's - 7th Infantry
David Brewer	Private	Dinkins 7th Infantry
William C. Sisson	Sergeant	Montgomery's, 7th Infantry
Reason Crump	Private	Montgomery, 7th Infantry
James O'Neal	Private	Montgomery's, 7th Infantry
Jackson Scarborough	Private	Montgomery's, 7th Infantry
Bannister Young	Corporal	Spotts' Detach., 7th Infantry
Wilson Wall	Private	Neilson's, 4th Infantry
Edward Deserne	Private	Muhlenberg's, 4th Infantry
Nathan Gorman	Private	Donoho's 4th Artillery
Unknown		Civilian, Female
Unknown		Civilian, Female
Unknown		Civilian, Female
Unknown		Civilian, Female

Unknown	Civilian, Female
Unknown	Civilian, Female
Unknown	Civilian, Child
Unknown	Civilian, Child
Unknown	Civilian, Child
Unknown	Civilian, Child

Total: 39

Wounded by Company

Name	*Rank*	*Unit*
William James	Private	Twiggs' Co., 7th Infantry
------ Gray	Unknown	Unknown
Unknown	Unknown	Unknown
Unknown	Unknown	Unknown
Unknown	Unknown	Unknown

Total: 5

Captured

Name	*Rank*	*Unit*
Elizabeth Stewart		Civilian, Female

Total: 1

Total Casualties

Killed	**39**
Wounded	**5**
Captured	**1**
TOTAL	**45**

Note: Native American casualties are unknown with the exception of three warriors captured and executed by Brig. Gen. William McIntosh who were reported to have been wounded in the attack on Scott's command.

Casualties

References

Arbuckle, Lt. Col. Matthew, to Brig. Gen. D. Parker and Staff, December 7, 1817, enclosed on the Monthly Report of the 7th U.S. Infantry for November 1817, Adjutant General's Office, Letters Received, 1805-1821.

Arbuckle, Lt. Col. Matthew, "Talk Delivered on the 10th of Decr. 1817 to three Indian Chiefs," December 10, 1817, transcribed by Captain Alex. Cummings, Office of the Adjutant General, Letters Received, 1805-1821.

Arbuckle, Lt. Col. Matthew, to Maj. Peter Muhlenberg, December 10, 1817, Office of the Adjutant General, Letters Received, 1805-1821.

Arbuthnot, Alexander, to the Commanding Officer at Fort Gaines, March 3, 1817, *American State Papers*, Indian Affairs, Volume II, p. 155.

Banks, Banks, Diary of John Banks, 1936.

Butler, Maj. Robert, Adjutant General, to Brig. Gen. Daniel Parker, Adjutant & Inspector General, May 3, 1818, *American State Papers*, Military Affairs, Volume 1, p. 703.

Calhoun, John C., Secretary of War, to Maj. Gen. Andrew Jackson, December 16, 1817, *American State Papers*, Indian Affairs, Volume 2, p. 162.

Calhoun, John C., Secretary of War, to Maj. Gen. Andrew Jackson, December 26, 1817, *American State Papers*, Indian Affairs, Volume 2, p. 162.

Clarke, Archibald, to Maj. Gen. E.P. Gaines, February 26, 1817, *American State Papers*, Indian Affairs, Volume II, p. 155.

Cook, Peter B. Cook to Elizabeth A. Carney, January 19, 1818, included in "Message of the President of the U. States to Congress, 25th March, 1818," published in the New York *Mercantile Advertiser*, January 6, 1819, p.2.

Dill, Col. John, to Gov. William Schley, May 27, 1836, Georgia State Archives.

Dill, Col. John, to Gov. William Schley, June 13, 1836, Georgia State Archives.

Dodd, Jordan, "Georgia Marriages to 1850 [database on-line]", Provo, UT, USA: Ancestry.com Operations Inc, 1997.

Executive Order dated Milledgeville, January 10, 1834, published in the *Macon Weekly Telegraph*, January 23, 1834, p. 3.

Executive Order dated Milledgeville, January 13, 1836, published in the *Macon Weekly Telegraph*, January 28, 1836, p. 3.

Ford, William P., to Gov. William Schley, May 26, 1836, Georgia State Archives.

Gadsden, Lt. James, to Maj. Gen. Andrew Jackson, May 3, 1818, *American State Papers*, Military Affairs, Volume 1, p. 715.

Gaines, Maj. Gen. Edmund P., to the Secretary of War, October 1, 1817, *American State Papers*, Indian Affairs, Volume II, pp. 158-159.

Gaines, Maj. Gen. Edmund P., to Maj. Peter Muhlenberg, October 11, 1817, Office of the Adjutant General, Letters Received, 1805-1821, National Archives.

Gaines, Maj. Gen. Edmund P., to the Commander of the Western Division, 8th Military District, October 18, 1817, Office of the Adjutant General, Letters Received, 1805-1821, National Archives.

Gaines, Maj. Gen. Edmund P., to Maj. Gen. Andrew Jackson, November 9, 1817, *American State Papers*, Indian Affairs, Volume II, p. 160.

References

Gaines, Maj. Gen. Edmund P., to Maj. Peter Muhlenberg, November 18, 1817, Office of the Adjutant General, Letters Received, 1805-1821, National Archives.

Gaines, Maj. Gen. Edmund P., to Maj. David E. Twiggs, November 20, 1817, Office of the Adjutant General, Letters Received, 1805-1821, National Archives.

Gaines, Maj. Gen. Edmund P., to Maj. Gen. Andrew Jackson, November 21, 1817, *American State Papers*, Indian Affairs, Volume II, p. 160.

Gaines, Maj. Gen. Edmund P., to Captain J.J. Clinch, November 30, 1817, Adjutant Generals Office, Letters Received, 1805-1821, National Archives.

Gaines, Maj. Gen. Edmund P., to Gov. William Rabun, December 2, 1817, published in the *Independent American*, January 14, 1818, p. 2.

Gaines, Maj. Gen. Edmund P., to the Secretary of War, December 2, 1817, *American State Papers*, Indian Affairs, Volume II, p. 687.

Gaines, Maj. Gen. Edmund P., General Orders of December 4, 1817, Office of the Adjutant General, Letters Received, 1805-1821.

Gaines, Maj. Gen. Edmund P., to the Secretary of War, December 4, 1817, *American State Papers*, Indian Affairs, Volume 2, p. 161.

Gaines, Maj. Gen. Edmund P., to the Secretary of War, December 15, 1817, *American State Papers*, Indian Affairs, Volume 2, p. 162.

Graham, George, Acting Secretary of War, to Maj. Gen. Edmund P. Gaines, October 30, 1817, *American State Papers*, Indian Affairs, Volume II, p. 159.

Graham, George, Acting Secretary of War, to Maj. Gen. Edmund P. Gaines, November 12, 1817, Office of the Adjutant General, Letters Received, 1805-1821.

Graham, George, Secretary of War, to Maj. Gen. Edmund P. Gaines, December 9, 1817, *American State Papers*, Indian Affairs, Volume 2, p. 161.

Hambly, William, and Edmund Doyle to "Sir," May 2, 1818, *American State Papers*, Military Affairs, Volume 1, p. 715.

Irvin, Capt. Robert, to Lt. Col. Matthew Arbuckle, December 23, 1817, *American State Papers*, Military Affairs, Volume 1, p. 692.

Jackson, Maj. Gen., to the Secretary of War, December 16, 1817, *American State Papers*, Indian Affairs, Volume 2, p. 162.

Jackson, Maj. Gen. Andrew, to John C. Calhoun, Secretary of War, March 15, 1818, *American State Papers*, Military Affairs, Volume 1, pp. 698-699.

Jackson, Maj. Gen. Andrew, to John C. Calhoun, Secretary of War, March 25, 1818, *American State Papers*, Military Affairs, Volume 1, pp. 698-699.

Jackson, Maj. Gen. Andrew, to John C. Calhoun, Secretary of War, April 8, 1818, *American State Papers*, Military Affairs, Volume 1, pp. 699-700.

Jackson, Maj. Gen. Andrew , to John C. Calhoun, Secretary of War, April 9, 1818, *American State Papers*, Military Affairs, Volume 1, p. 700.

Jackson, Maj. Gen. Andrew, to John C. Calhoun, Secretary of War, April 20, 1818, *American State Papers*, Military Affairs, Volume 1, pp. 700-701.

Jackson, Maj. Gen. Andrew, to John C. Calhoun, Secretary of War, May 5, 1818, *American State Papers*, Military Affairs, Volume 1, pp. 701-702.

Johnson, Lt. Milo, to Maj. Gen. Edmund P. Gaines, November 30, 1817, Office of the Adjutant General, Letters Received, 1805-1821, National Archives.

Cappachimico to Commanding Officer at Fort Hawkins, September 18, 1817, *American State Papers*, Indian Affairs, Volume II, p. 159.

References

McAdams, Eileen Babb, "Georgia Militia 1847, Major and Brigadier Generals," (List of officers assembled from Augusta Chronicle, March 31, 1848), U.S. GenWeb.

McIntosh, Capt. John N. to Hon. A. Lacock, February 5, 1819, *American State Papers*, Military Affairs, Volume 1, p. 747.

McIntosh, Brig. Gen. William, to Maj. Daniel Hughes, March 2, 1818, *Camden Gazette*, April 11, 1818, p. 3.

McIntosh, Brig. Gen. William, to Maj. Daniel Hughes, March 5, 1818, *Camden Gazette*, April 11, 1818, p. 3.

McIntosh, Brig. Gen. William, to Maj. Daniel Hughes, March 10 & 16, 1818, *Camden Gazette*, April 11, 1818, p. 3.

McIntosh, Brig. Gen. William, to David B. Mitchell, Agent for Indian Affairs, April 13, 1818, *Berks and Schuylkill Journal*, May 16, 1818, p. 2.

Minutes of the Trials of Alexander Arbuthnot and Robert C. Ambrister, April 26 & 27, 1818, *American State Papers*, Military Affairs, Volume 1, pp. 721-735.

Mitchell, David B., Agent for Indian Affairs, to the Secretary of War, March 30, 1817, *American State Papers*, Indian Affairs, Volume II, pp. 156-157.

Mitchell, David B., Agent for Indian Affairs, to Acting Secretary of War George Graham, December 14, 1817, *American State Papers*, Indian Affairs, Volume 2, p. 161.

Neamathla, Speech of Neamathla, September 11, 1823, *American State Papers*, Indian Affairs, Volume II, p. 439.

Newspapers:
o *Berks and Schuylkill Journal*, May 16, 1818.
o *Camden Gazette*, April 11, 1818.
o *Georgia Journal*, July 24, 1830.
o *Hampshire Gazette*, November 26, 1817.

- *Independent American*, January 14, 1818.
- *Massachusetts Spy*, December 31, 1817.
- *Mercantile Advertiser*, January 6, 1819.
- *Milledgeville Reflector*, December 9, 1817.
- *Macon Weekly Telegraph*, January 23 & 28, 1834.
- *Poulson's American Daily Advertiser*, July 11, 1817.
- *Providence Patriot*, November 29, 1817.
- *Rutland Herald*, January 7, 1818.
- *Savannah Republican*, December 17, 1817.
- *Vermont Reporter*, May 20, 1817.
- *Weekly Aurora*, May 25, 1818 & June 22, 1818.

Officer (Unknown), Letter from an officer at Fort Scott to his father, December 2, 1817, *Massachusetts Spy*, December 31, 1817, p. 2.

Original Land Lot Surveys of Early County, Georgia, 1819-1820, Georgia State Archives.

Patterson, J., A. McGinty and J. Dill to Gov. William Schley, January 29, 1836, Georgia State Archives.

Perryman, George, to Lt. R. Sands, February 24, 1817, *American State Papers*, Indian Affairs, Volume II, p. 155.

"Register for Details for Command from Fort Scott, from the 18th of December, 1817, until the 19th of March, 1818, whilst under the command of Lt. Col. Arbuckle," Office of the Adjutant General, Letters Received, 1805-1821.

Sands, Lt. Richard M., to Col. William King, March 15, 1817, *American State Papers*, Indian Affairs, Volume II, p. 156.

Scott, Lt. Richard W., to Maj. Gen. Edmund P. Gaines, November 28, 1817, *American State Papers*, Indian Affairs, Volume II, p. 688.

Service Record of Private John Dill, State Militia Records, South Carolina State Archives.

References

Stuart-Purcell Map of 1778, Cartographic Division, Library of Congress.

Tustennogee Hopoi and Hopoi Haija to David B. Mitchell, Indian Agent, December 30, 1817, *American State Papers*, Military Affairs, Volume 1, pp. 692-693.

Twiggs, Maj. David E., to Maj. Gen. Edmund P. Gaines, September 17, 1817, *American State Papers*, Indian Affairs, Volume II, p. 158.

Twiggs, Maj. David E., to Maj. Gen. Edmund P. Gaines, November 21, 1817, Office of the Adjutant General, Letters Received, 1805-1821, National Archives.

U.S. Army, Post Returns, Fort Scott, Georgia, 1817-1821, National Archives.

U.S. Army, Register of Enlistments, 1798-1815, National Archives.

U.S. Census for District 749, Clay County, Georgia, 1860.

Woodward, Gen. Thomas S., to Col. John Banks, June 16, 1858, *Woodward's Reminiscences of the Creek, or Muscogee Indians, Containted in Letters to Friends in Georgia and Alabama*, Montgomery, Alabama, 1859.

Wright, Maj. Clinton, Assistant Adjutant General, to Maj. Peter Muhlenberg, December 2, 1817, Office of the Adjutant General, Letters Received, 1805-1821, National Archives.

Young, Capt. Hugh, "A Topographical Memoir of East and West Florida with Itineraries," 1818, National Archives.

Index

Index

Darby
William, 169
Department of Texas, 144
Deserne
Edward, 169
Dill
Elizabeth Stewart. *See* Elizabeth
Stewart
John, 135, 136, 137, 139, 140, 163
District of Louisiana, 144
Donoho
Capt. Samuel, 12, 24, 25, 76
Dothan, 150
Doyle
Edmund, 18, 44, 45, 48, 69, 74, 89,
94, 131
Driver
Jonathan, 169
Duval
Gov. William P., 144
Earthquake of 1811, 2
Econchattimico, *92*, 144
Edwards
Corp. James, 169
Ekanachatte, 16, *91*, *92*, 117
Ellicott
Andrew, 11
Emathla
Charley, 149
Eneah Emathla. *See* Neamathla
Eneamathla. *See* Neamathla
E-nee-hee-maut-by. *See* Neamathla
Eufaula, 139, 145
Execution
Arbuthnor & Ambrister, 132
Fort Gaines, *91*, 155
Francis & Homathlemico, *101*, *102*,
104, *105*
Federal Road, 21
First Brigade, 21, 23, 24
First Seminole War, 51, 52, 55, *103*,
133, 137
Five Civilized Tribes, 148
Flint River, 1, 2, 3, 4, 7, 12, 15, 16, 17,
19, 21, 22, 23, 24, 26, 28, 30, 45, 48
Florida State University, 149
Floyd
Maj. Gen. John, *103*

Forbes & Company, 18, 44
Fort Apalachicola, 71
Fort Early, *91*, *93*, *95*
Fort Gadsden, *96*, *97*, 133, 147
Fort Gaines, 3, 7, 10, 11, 21, 23, 24, *90*,
91, 117, 135, 136, 137, 138, 139,
150, 155
Fort Gaines Guards, 140
Fort Gibson, 146
Fort Hawkins, 17, 24, 79, 82, 83, *90*,
152
Fort Hughes, 30, 42, 74, 76, 90
Fort King, 149
Fort Marion. *See* Castillo de San Marcos
Fort Mims, *103*
Fort Mitchell, 3, 21, 81, 155
Fort Montgomery, 19, 24, 153
Fort Moultrie, 149
Fort Peyton, 149
Fort Scott, 1, 3, 5, 11, 12, 13, 17, 19, 21,
22, 23, 24, 25, 26, 27, 28, 29, 30, 43,
44, 45, 48, 54, 58, 67, 69, 70, 73, 79,
81, 85, 90, *93*, *94*, *97*, *101*, 146, 151,
156
Fort Smith, 136, 146
Fort St. Marks, *94*, *102*, *104*, 108, 130,
133, 147
Fort Twiggs, 139
Four Mile Creek, 15, 16
Fowltown, 3, 4, 5, 7, 15, 16, 19, 20, 25,
26, 28, 30, 40, 43, 44, 52, 81, 83, 86,
89, *101*, 133, 143, 153
Fowltown Swamp, 26, 39
Francis
Josiah, 3, 4, 45, 53, *94*, *101*, *102*,
104, 129, 131, 147
Milly, *104*, 123, 147, 165
French and Indian War, 15
Gadsden
Lt. James, *96*, 130
Gadsden Purchase, *96*
Gaines
Maj. Gen. Edmund P., 19, 20, 54, 58,
66, 72, 79, 82, 135, 145, 153
Maj. Gen. Edmund Pendleton, 9, 10,
12, 21, 22, 23, 24, 25, 26, 27, 28,
43, 45, 47, 48
Maj. Gen. Edmund Pendlton, 48

183

Index

185

About the Author

Dale Cox is a writer and historian with deep roots in Northwest Florida, Southwest Georgia and Southeast Alabama where the Seminole Wars began. He has drawn critical acclaim for his books on military history.

The author's other books about the Creek War/Seminole War era include *Milly Francis: The Life & Times of the Creek Pocahontas, Fowltown: Neamathla, Tutalosi Talofa, and the First Battle of the Seminole Wars, Fort Scott Fort Hughes, and Camp Recovery,* and *Fort Gaines, Georgia: A Military History.*

He has also written books about the Battles of Marianna and Natural Bridge in Florida, the Battle of Massard Prairie in Arkansas, and the infamous 1934 lynching of Claude Neal in Jackson County, Florida. On a lighter side, his volumes about his home community of Two Egg, Florida, have achieved widespread popularity and his short novel, *A Christmas in Two Egg, Florida*, was adapted for the stage.

A descendant of both American frontiersman Daniel Boone and 19th-century Yuchi Indian leaders, Cox has a love for history that has translated to involvement in numerous historic preservation efforts. He is a co-founder of the streaming travel and history channel Two Egg TV and is a popular speaker for groups and events across the Southeast.

Dale Cox is the author of two grown sons, William and Alan. He is a Christian and resides near Two Egg in the piney woods of Northwest Florida with his magnificent Huskie, Dodger D. Dogg.

Books by Dale Cox

Fowltown
Neamathla, Tutalosi Talofa, and the First Battle of the Seminole Wars

Fort Gaines, Georgia
A Military History

Fort Scott, Fort Hughes and Camp Recovery
Three 19th Century Military Sites in Southwest Georgia

Milly Francis
The Life & Times of the Creek Pocahontas

The Battle of Natural Bridge, Florida
The Confederate Defense of Tallahassee

The Battle of Massard Prairie
The 1864 Confederate Attacks on Fort Smith, Arkansas

The Claude Neal Lynching
The 1934 Murders of Claude Neal and Lola Cannady

Two Egg, Florida
A Collection of Ghost Stories, Legends & Unusual Facts

A Christmas in Two Egg, Florida
A Short Novel of Redemption

The Ghost of Bellamy Bridge
10 Ghosts & Monsters from Jackson County, Florida

Books by Dale Cox can be ordered online at www.exploresouthernhistory.com
and also are available as instant downloads for Kindle users.

www.ingramcontent.com/pod-product-compliance
Lightning Source LLC
Chambersburg PA
CBHW021400090426
42742CB00009B/936